# Sourcebook for

# Electronics Calculations, Formulas, and Tables

# Sourcebook for
# Electronics Calculations, Formulas, and Tables

### Newton C. Braga

PROMPT®
PUBLICATIONS

**PROMPT© Publications** is an imprint of Howard W. Sams & Company, A Bell Atlantic Company, 2647 Waterfront Parkway, E. Dr., Indianapolis, IN 46214-2041.

**International Standard Book Number: 0-7906-1193-7**
**Library of Congress Catalog Card Number: 99-63846**

*Acquisitions Editor*: Loretta Yates
*Editor*: Pat Brady
*Assistant Editor*: J.B. Hall
*Typesetting*: Pat Brady
*Cover Design*: Christy Pierce
*Graphics Conversion*: Dick Raus, Bill Skinner, Terry Varvel
*Illustrations and Other Materials*: Courtesy of the Author

PRINTED IN THE UNITED STATES OF AMERICA

9 8 7 6 5 4 3 2 1

# Contents

PREFACE ................................................................ 1

## Part 1. DC Formulas ................................ 5

1. UNITS ................................................................ 7
   1.1 - Electric Units ............................................ 7
   TABLE 1. Basic Electric Units and Symbols: ...................... 8
   TABLE 2. Metric prefixes used in the SI to express multiples and submultiples of the units above (and others): ........................ 9
   TABLE 3. Conversion Table and Applied Electric Units: ............ 10
   TABLE 4. Conversion Table for Some Non-electric Important Units: ....... 12
   TABLE 5. Numerical Latin/Greek Prefixes ......................... 14

2. RESISTANCE OF A CONDUCTOR .................................... 16
   TABLE 6. Specific Resistance Of Some Materials .................. 17

3. CONDUCTANCE (I) .............................................. 18

4. CONDUCTANCE OF A LENGTH OF WIRE .............................. 19
   TABLE 7. Conductance of Some Common Materials .................. 22

5. THERMAL INFLUENCE IN RESISTANCE OF A CABLE .................. 23
   TABLE 8. Temperature Coefficient of Resistivity of Materials at 20oC ....... 23
   TABLE 9. Standard Annealed Copper Wire (AWG & B&S) ............. 24

5. OHM'S LAW .................................................... 26

6. POWER ....................................................... 27

7. JOULE'S LAW ................................................. 28

8. ELECTRIC ENERGY ............................................. 30

9. ELECTROLYSIS (Faraday's Law) ......................................... 32
    TABLE 10. Electrochemical Equivalents of Some Substances ................. 32

10. RESISTORS IN SERIES ................................................. 34

11. RESISTORS IN PARALLEL ............................................... 35
    Table 11. Color Code for Resistors ...................................... 38

12. VOLTAGE DIVIDER ..................................................... 39

13. LOADED VOLTAGE DIVIDER ............................................. 40

14. FIRST KIRCHHOFF'S LAW .............................................. 41

15. SECOND KIRCHHOFF'S LAW ............................................ 42

16. CAPACITANCE ........................................................ 43

17. PLANAR CAPACITOR ................................................... 44
    TABLE 12. Dielectric Constant for Some Materials ....................... 46

18. BREAKDOWN VOLTAGE IN A CAPACITOR ................................ 47
    TABLE 13. Dielectric Strength of Some Materials ........................ 49

19. ENERGY STORED IN A CAPACITOR ...................................... 50

20. CAPACITORS IN PARALLEL ............................................. 52

21. CAPACITORS IN SERIES ............................................... 53

22. MAGNETIC FIELD IN A SOLENOID ....................................... 55

23. MAGNETIC INDUCTION INSIDE A SOLENOID ............................. 57

24. INDUCTANCE ......................................................... 58

25. INDUCTANCES IN SERIES ............................................. 60

26. INDUCTANCES IN PARALLEL ........................................................ 61

27. MUTUAL INDUCTANCE ............................................................... 62

## Part 2.  AC Formulas ........................................................ 65

28. FREQUENCY AND PERIOD.......................................................... 67

29. CYCLIC OR ANGULAR FREQUENCY ........................................... 68

30. AVERAGE VALUE ...................................................................... 69

31. RMS VALUE ............................................................................... 71
TABLE 15. Conversion Factors or Multiplier for Converting Maximum, Average and rms Values ......................................................................................... 72

32. FREQUENCY & WAVELENGTH .................................................... 72
TABLE 16. Velocity of Sound in Solids at 20°C ........................................ 73
TABLE 17. Velocity of Sound in Gas (1 atm x 0°C) ................................. 74

33. CAPACITIVE REACTANCE ......................................................... 75

34. INDUCTIVE  REACTANCE .......................................................... 77

35. QUALITY FACTOR (FACTOR-Q) .................................................. 79

36. OHM'S LAW FOR AC CIRCUITS .................................................. 81

37. RL IN SERIES ............................................................................ 82

38. RC IN SERIES ........................................................................... 84

39. LC IN SERIES ............................................................................ 85

40. RLC IN SERIES .......................................................................... 87

41. RC IN PARALLEL........................................................................ 88

42. LR IN PARALLEL ............................................................. 89

43. LC IN PARALLEL ............................................................. 90

44. RLC IN PARALLEL ........................................................... 91

45. RESONANCE (LC) ........................................................... 92

46. TIME CONSTANT (RC circuit) ........................................ 95
    TABLE 19. Derivated Formulas ........................................ 96

47. TIME CONSTANT (LC) ..................................................... 97

48. INDUCTIVE COUPLING USING TRANSFORMERS ............. 99

49. DIRECT INDUCTIVE COUPLING ...................................... 100

50. OHMIC COUPLING .......................................................... 100

51. CAPACITIVE COUPLING .................................................. 101

52. LOW-PASS FILTERS ...................................................... 101

53. HIGH-PASS FILTERS ..................................................... 103

54. BAND-PASS FILTERS .................................................... 104

55. DIFFERENTIATION ......................................................... 106

56. INTEGRATION ................................................................ 107

57. NOISE ........................................................................... 107

58. BANDWIDTH .................................................................. 108

59. VOLTAGE RATIO IN TRANSFORMERS .............................. 109

# Contents

60. CURRENT RATIO IN TRANSFORMERS .................................................... 110
    TABLE 20 Impedance Ratio and Turns Ratio of Transformers ............................ 112

61. DECIBEL ........................................................................................... 113
    TABLE 21 Voltage or Current Ratios vs Ratios and Decibels ........................... 115

62. The Neper ......................................................................................... 117
    TABLE 22 Relationships Between Decibels and Nepers .................................. 117

63. BALANCED T-ATTENUATOR .................................................................. 117

64. BALANCED Pi-ATTENUATOR .................................................................. 119

65. UNBALANCED T-ATTENUATOR ............................................................... 121

66. UNBALANCED Pi-ATTENUATOR ............................................................... 122

67. HALF-WAVE DIPOLE .............................................................................. 124

68. FOLDED HALF-WAVE DIPOLE .................................................................. 125
    TABLE 23. Frequency x Wavelength for Some Frequencies ............................ 126

69. RANGE .............................................................................................. 127

70. COAXIAL CABLE .................................................................................. 128

71. TWO-WIRE BALANCED LINE .................................................................. 129

72. IMPEDANCE MATCH - Pi-Network ........................................................... 131

## Part 3.  Electronic Circuits ........................................... 133

73. SEMICONDUCTOR DIODE ...................................................................... 135

74. HALF-WAVE RECTIFIER .......................................................................... 135

75. FULL-WAVE RECTIFIER .......................................................................... 137

76. LC FILTER COEFFICIENT ........................................................ 139

77. RC FILTER COEFFICIENT ........................................................ 141

78. RIPPLE FACTOR ................................................................... 142
    TABLE 24. Rectifier's Characteristics (using resistive loads) ...................143

79. FILTER INDUCTANCE ............................................................ 144

80. FILTER CAPACITANCE .......................................................... 145

81. CONVENTIONAL VOLTAGE DOUBLER ..................................... 146

82. CASCADE VOLTAGE DOUBLER ............................................. 146

83. BRIDGE VOLTAGE DOUBLER ................................................ 147

84. FULL WAVE TRIPLER .......................................................... 148

85. CASCADE VOLTAGE TRIPLER ............................................... 149

86. FULL-WAVE VOLTAGE QUADRUPLER ..................................... 150

87. ZENER DIODE .................................................................... 151

88. CAPACITIVE VOLTAGE DIVIDER ............................................ 154

89. NTC ................................................................................. 157

90. PTC ................................................................................. 159

91. VARICAPS ........................................................................ 160

**TRANSISTORS ..................................................... 161**

92. TRANSISTOR STATIC CURRENT-GAIN (Common Emitter) ....... 161

93. TRANSISTOR STATIC CURRENT GAIN (Common-Base Configuration) .. 162

94. RELATIONSHIP BETWEEN ALPHA AND BETA ......................................... 163

    TABLE 25. Alpha & Beta Conversion ................................................................ 164

95. HYBRID PARAMETERS .............................................................................. 165

    TABLE 26. Letter Symbols and Abbreviations for Transistors ......................... 168

96. COMMON-BASE ......................................................................................... 168

97. COMMON EMITTER ................................................................................... 169

98. COMMON-COLLECTOR ............................................................................. 170

    TABLE 27. General Characteristics of Transistor Circuits ................................ 171

# BASIC QUANTITIES OF CIRCUITS USING TRANSISTORS ........ 171

99. SHORT-CIRCUIT OUTPUT .......................................................................... 172

100. OPEN-CIRCUIT OUTPUT .......................................................................... 173

101. SHORT-CIRCUIT INPUT ............................................................................ 174

102. OPEN-CIRCUIT INPUT .............................................................................. 175

103. COMMON-BASE CONFIGURATION USUAL FORMULAS ....................... 176

104. COMMON-EMITTER CONFIGURATION USUAL FORMULAS ................... 180

105. COMMON-COLLECTOR USUAL FORMULAS ........................................... 182

    TABLE 28. Other Symbols used in Transistor Specifications ........................... 185

    TABLE 29. Other Symbols for Common-Emitter Configuration ......................... 185

    TABLE 30. Other Symbols for Common-Base Configuration ............................ 186

    TABLE 31. Other Symbols for Common-Collector Configuration ..................... 186

# TRANSISTOR PRACTICAL FORMULAS ..................................... 186

106. LOAD RESISTANCE ................................................................................. 187

107. BASE-BIASING RESISTANCE .................................................................. 188

108. AUTOMATIC-BIASING RESISTANCE ...................................................... 190

## JUNCTION FIELD-EFFECT TRANSISTOR (JFET) AND MOS FIELD-EFFECT TRANSISTOR (MOSFET) ......................................... 191

109. COMMON SOURCE ...................................................................... 191

110. COMMON-DRAIN ........................................................................ 193

111. COMMON-GATE .......................................................................... 195

112. UNIJUNCTION TRANSISTOR (UJT) ................................................. 196

113. SCR .......................................................................................... 200

114. TRIAC ....................................................................................... 203

## OSCILLATORS ................................................................................ 205

115. ASTABLE MULTIVIBRATOR ........................................................... 205

116. NEON-LAMP OSCILLATOR ............................................................ 208

117. PHASE-SHIFT OSCILLATOR ...........................................................211

118. Wien Bridge Oscillator ............................................................... 213
    TABLE 32. Greek Alphabet ............................................................ 216
    TABLE 33. Equally Tempered Chromatic Scale Frequencies (Hz) ......... 218

120. HARTLEY OSCILLATOR ............................................................... 218

121. COLPITTS OSCILLATOR ............................................................... 219

122. CMOS TWO-GATE OSCILLATOR (I) ............................................... 221
    TABLE 34. CMOS ICs Suitable for Using as Oscillators: .................... 225

124. CMOS SCHMITT TRIGGER OSCILLATOR ....................................... 225
    TABLE 35. Threshold Voltages of the 4093 CMOS IC ...................... 227

125. THE ASTABLE 555 .......................................................... 227
    TABLE 36. Limit Values Recommended For the Astable 555 ............... 230
    GRAPH.  1555 Free Running Frequency ................................. 231

126. MONOSTABLE 555 ........................................................... 232
    TABLE 37. Limit Values for the Monostable 555 ...................... 233
    GRAPH.  2555 Monostable - On time versus R and C ................... 234

# BRIDGES ................................................................... 235

127. WHEATSTONE BRIDGE ...................................................... 235

128. WIEN BRIDGE ............................................................. 236

129. RESONANCE BRIDGE ....................................................... 238

130. MAXWELL BRIDGE ......................................................... 239

131. SCHERING BRIDGE ........................................................ 241

132. OWEN BRIDGE ............................................................. 243

133. HAY BRIDGE .............................................................. 244

# OPERATIONAL AMPLIFIERS .................................................... 245

134. NONINVERTING OP AMP .................................................... 245

135. INVERTING OP AMP ....................................................... 246

# VOLTAGE FOLLOWER .......................................................... 247

136. SUMMING OP AMP .......................................................... 248

137. SUBTRACTION OP AMP ..................................................... 249
    TABLE 38. CMRR vs dB ................................................ 250

138. DIFFERENTIATION USING OP AMP .......................................... 251

139. INTEGRATION USING OP AMP ............................................................... 252

140. LOGARITHMIC OP AMP .................................................................... 253

141. VOLTAGE SOURCE .......................................................................... 254

142. CONSTANT CURRENT SOURCE - Op Amp Floating Load ...................... 256

143. CONSTANT CURRENT SOURCE - High Current Load ......................... 257

144. ABSOLUTE VALUE AMPLIFIER - Op Amp ......................................... 257

145. VOLTMETER USING Op Amp ............................................................ 259

146. SQUARE WAVE OSCILLATOR - Op Amp ........................................... 259

147. WIEN BRIDGE OSCILLATOR using Op Amp ...................................... 261

148. BANDPASS AMPLIFIER using Op Amp ............................................. 262

149. NOTCH FILTER - Op Amp ................................................................ 264

150. LOW-PASS FILTER - Op Amp .......................................................... 265

151. HIGH-PASS FILTER - Op Amp ......................................................... 266

152. BUTTERWORTH FILTER - Op Amp ................................................... 267

153. SERIES VOLTAGE REGULATOR (one transistor) ................................. 268
    TABLE 39. Common Values of Expressions using p ............................. 275

154. PARALLEL VOLTAGE REGULATOR .................................................. 276

**INTEGRATED VOLTAGE REGULATORS ..................................... 280**

155. VOLTAGE REGULATORS ................................................................. 281

156. CURRENT REGULATOR ............................................................ 283
    TABLE 403. Terminal Voltage Regulator ICs ..................................... 285

# Part 4.  Digital ........................................................ 287

157. BINARY-TO-DECIMAL CONVERSION ................................. 289
    TABLE 41. Powers of Two ................................................................ 290

158. BYTE-TO-DECIMAL CONVERSION .................................... 291
    TABLE 42. Decimal Integers to Pure Binaries ................................. 291

159. BCD TO DECIMAL ............................................................ 292
    TABLE 43. Negative Powers of Two ................................................. 294

160. HEXADECIMAL-TO-DECIMAL CONVERSION ..................... 294
    TABLE 44. Hexadecimal Digits and Decimal Correspondents ........... 296
    TABLE 45. Powers of 16 .................................................................. 296

161. DECIMAL-TO-BINARY CONVERSION ............................... 296

# LOGIC FUNCTIONS ............................................................ 297

162. AND GATE ....................................................................... 298
    TABLE 46. Truth Table - 2-Input AND Function .............................. 299
    TABLE 47. Truth Table - 3-Input NAND Gate .................................. 299

163. NAND GATE ..................................................................... 300
    TABLE 48. Truth Table — 2-Input NAND Gate ................................ 301
    TABLE 49. Truth Table — 3-Input NAND Gate ................................ 301

164. OR GATE ......................................................................... 302
    TABLE 50. Truth Table — 2-Input OR Gate .................................... 303
    TABLE 51. Truth Table — 3-Input OR Gate .................................... 303
    TABLE 52. Truth Table — 2-Input NOR Gate .................................. 305
    TABLE 53. Truth Table — 3-Input NOR Gate .................................. 305

166. EXCLUSIVE-OR ............................................................... 306
    TABLE 54. Truth Table — Exclusive-OR Gate ................................. 306

168. BINARY ADDITION ............................................................. 307

169. BINARY SUBTRACTION .................................................... 308

170. BINARY MULTIPLICATION ............................................... 308

171. BINARY DIVISION ............................................................ 308

## THE POSTULATES OF BOOLEAN ALGEBRA ............................ 309

172. LAWS OF TAUTOLOGY .................................................... 309

173. LAWS OF COMMUTATION............................................... 310

174. LAWS OF ASSOCIATION ..................................................311

175. LAWS OF DISTRIBUTION ................................................ 312

176. LAWS OF ABSORPTION .................................................. 313

177. LAWS OF UNIVERSE CLASS ........................................... 314

178. LAWS OF NULL CLASS .................................................... 315

179. LAWS OF COMPLEMENTATION ...................................... 316

180. LAWS OF CONTRAPOSITION ......................................... 317

181. LAW OF DOUBLE NEGATION ......................................... 318

182. LAWS OF EXPANSION..................................................... 319

183. LAWS OF DUALITY ......................................................... 320

## BOOLEAN RELATIONSHIPS ................................................. 321

184. IDEMPOINT ................................................................... 321

185. COMMUTATIVE ............................................................. 322

186. ASSOCIATIVE ............................................................... 322

187. DISTRIBUTIVE .............................................................. 322

189. DEMORGAN THEOREM ............................................... 323
    TABLE 55. Some TTL AND/NAND Gates ........................... 323
    TABLE 56. Common TTL OR/NOR Gates ........................... 324
    TABLE 57. TTL Subfamily Fan-Out Rules ........................... 324

**Part 5. Miscellaneous** ..................................... **327**

190. INSTANTANEOUS VALUES OF SINE WAVES ................. 329
    TABLE 58. Instantaneous Values of Sine Wave (Current or Voltage) ................... 329

191. INCHES TO MILLIMETERS ........................................... 333
    TABLE 59. Decimal Inches to Millimeters ........................... 335
    TABLE 60. Millimeters to Decimal Inches ........................... 336

**TEMPERATURE CONVERSIONS** ............................... **337**

192. DEGREES CELSIUS (C), DEGREES FAHRENHEIT (F), DEGREES KELVIN (K) AND DEGREES Kelvin to Fahrenheit ................... 339
    TABLE 61. Degrees Celsius to Fahrenheit Conversion ......... 341
    TABLE 62. Farenheit to Celsius Conversion ........................ 342

**SOUND** ............................................................... **342**

193. SOUND WAVELENGTH ................................................ 342

194. VELOCITY OF LONGITUDINAL SOUND WAVES ............. 344
    TABLE 63. Young's Modulus for some Solid Materials ......... 345

195. SOUND PRESSURE ..................................................... 346

196. SOUND PRESSURE LEVEL ........................................... 347

197. SOUND INTENSITY ....................................................... 348
    TABLE 64. Sound Intensity Levels ................................. 349

## LOUDSPEAKER CROSSOVER NETWORKS ............................. 350

198. SIMPLE FOR TWEETER (6 db/octave) .................................... 350

199. 6 dB/Octave CROSSOVER NETWORK-II ................................... 351

200. TWO-CHANNEL 12 dB LOUDSPEAKER CROSSOVER .......................... 353

201. TWO-CHANNEL 18 dB/octave PI-CROSSOVER ............................. 354

202. TWO-CHANNEL 18 dB/Octave T-CROSSOVER .............................. 356

203. CROSSOVER AND DESIGN FREQUENCIES ..................................
    FOR 3-WAY NETWORKS: ................................................. 358

204. THREE-CHANNEL 6dB/Octave CROSSOVER  (Series) ...................... 360

205. THREE-CHANNEL 6 dB/Octave CROSSOVER (Parallel) .................... 361

206. THREE-CHANNEL 12 dB/octave CROSSOVER (Series) ..................... 363

207. THREE-CHANNEL 12 dB/Octave CROSSOVER (Parallel) ................... 365

208. WEELER'S FORMULA .................................................... 367

## Part 6. Optoelectronics .................................. 371

    TABLE 65. Optical Radiation and Light ............................ 373

209. WAVELENGTH AND FREQUENCY FOR LIGHT RADIATION ................. 373
    TABLE 66. Photometric Conversion Table .......................... 375

210. LUMINOUS INTENSITY AND FLUX ....................................... 376

211. LUMINOUS DENSITY ......................................................... 377
  TABLE 67. Radiometric and Photometric Parameters ..................... 379
  TABLE 68. Spectral Units .............................................. 380
  TABLE 69. Photopic Sensitivity of the Eye ............................. 380
  TABLE 70. Characteristic of Some Common Light Sources ................. 382

212. LEDs ....................................................................... 382
  TABLE 71. Forward Voltage Fall (Uf) in LEDs .......................... 383

213. PHOTOCELL ................................................................. 384

214. LDR OR PHOTORESISTOR .................................................... 386

215. PHOTODIODE ............................................................... 389

216. PHOTOTRANSISTOR ......................................................... 390

**COLORIMETRY** ............................................................... **392**

  TABLE 72. Color Primary Valences ..................................... 392
  TABLE 73. Standard Color Valences .................................... 392

217. STEFAN-BOLTZMANN'S LAW ................................................. 392

**MATHEMATICS** ............................................................... **394**

218. DIFFERENTIATION FORMULAS .............................................. 394

219. INTEGRATION FORMULAS ................................................... 396
  TABLE 74. Square and Cubic Roots of Some Numbers ..................... 398

220. TOLERANCES ............................................................... 401
  TABLE 75. Laplace Transforms ......................................... 403

221. FOURIER TRANSFORMS ..................................................... 405
  TABLE 76. Square Wave Harmonic Composition ........................... 406
  TABLE 77. Triangular Wave Harmonic Composition ....................... 407
  TABLE 78. Sawtooth Wave Harmonic Composition ......................... 408

TABLE 79. Half-Wave Rectifier Waveform Harmonic Composition ........................ 409

TABLE 80. Full-Wave Rectifier Waveform Harmonic Composition ........................ 410

TABLE 81. Physical Constants ............................................................................ 411

# REFERENCES ................................................. 413

# PREFACE

This was written for anyone who designs electronic circuits. Engineers, technicians, students and hobbyists can really benefit from this book. The author, who has himself designed multitudes of projects and circuits during his life, publishing many books and hundreds of articles in electronics magazines, has collected an assortment of all basic information necessary for calculations needed when designing new projects.

When starting a project, the principal difficulty the designer finds is how to locate the desired information. This information is normally spread over a large number of resources, such as books, handbooks and magazine articles.

Although many of us who are experienced in electronics have in mind the principal formulas, we sometimes have trouble with the forgotten constant, multiplication factor or exponent. Finding these values is sometimes difficult depending of the circumstances, such as where you are at the time, or the amount of resources at your disposal.

By putting the principal formulas and tables in a unique place, a designer can find the desired information easier, and, more importantly, can take this information wherever he goes. This is the aim of this book.

But formulas and tables are not useful only when designing a new configuration. They are necessary when we need to know what will happen when a specific working circuit is altered, for the electronic student doing homework, or the researchers in other fields who work with electronic equipment.

The tables contain a large amount of important information, such as particular values of constants, physical properties of circuits and materials, and even calculated values that can't be found without using complex or hard-to-do procedures.

Finally, we have laws and theorems describing the properties of circuits and devices, and procedures to be used in calculations, which are very important when doing practical works.

Most of the formulas and tables are accompanied by application examples. They are very important to show the reader how the calculations are made when using the given information. To avoid problems with incorrect results, in all formulas and applications the units to be used are indicated.

The formulas range from the simplest, where elementary arithmetic operations such as sum, subtraction, multiplication and division, are used, to the more complex that require some good working knowledge of algebraic and trigonometric functions, or even integral and differential calculus.

Although mathematics is an exact science, when some calculations are applied to other sciences and in "real-life" electronics, the results can be different from expected. When making calculations involving electronic circuits, it is often said that "When working with electronics, practice and theory often disagree."

This means that in many cases the results found in calculations will need some "adjustments" when applied to actual application.

This fact is applicable even to the tolerances of the electronic components used in practical applications, plus the fact that many formulas are not exact, but empirical.

But why use an exact formula, including complex logarithms, trigonometric functions or differential equations, if we can get results good enough to make a circuit work by using a reduced formula?

In many stages of the design process, the results will depend on the tolerances of the components used.

This explains why in many cases we'll not give the exact formula but an empirical formula, where the "complex part" of the calculations will be reduced to a constant, or even eliminated.

About the units—the preference is the use of the decimal system or SI. Only in the cases where conversion formulas and tables are given will other units appear. The notations will be that recommended by NIST (National Institute of Standards and Technology), but in some cases, to make easier the use by readers less experienced with calculations, some "nonconventional" notations can be found.

Although the preferred symbol to indicate multiplication operation is "x," in some cases we also use the bullet (•).

The tables were obtained from different sources—physics handbooks, engineering books and manufacturers' information were consulted. As for composition of materials, the calculation procedures can change from one manufacturer to another, so small differences in the tabulated values will be found if comparing to other resources.

Many tables were calculated by the author using computer programs created to do the task.

Using the book is very easy. The formulas are distributed in groups according to subject, and all the reader needs is a calculator and a piece of paper to work. A scientific calculator is recommended if trigonometric functions or logarithms are used.

The most important point of this introduction is that this sourcebook can become an important tool for everyone who works with electronic design—from students and hobbyists to researchers and engineers. Anyone in need of practical information on formulas and calculations as applied to electronics will find this book indispensable.

Newton C. Braga

# Part 1


# DC Formulas

# 1. UNITS

## 1.1 - Electric Units

In the next pages we'll give the reader some of the most used symbols and units for electric units. In many cases the metric prefixes are included. The units are in alphabetical order to make easier to the reader find the corresponding symbol.

All the units are based in the International System of Units (*Système International d'Únités*), or SI, adopted in the *Conférence Générale des Poids et Mesures* in 1960.

This system used as basic units the following:

| Quantity | Unit | Symbol |
|---|---|---|
| length | meter | m |
| mass | kilogram | kg |
| time | second | s |
| electric current | ampere | A |
| temperature | kelvin | K |
| luminous intensity | candela | cd |
| amount of substance | mole | mol |

## TABLE 1

## Basic Electric Units and Symbols:

| Unit | Unit Symbol | Quantity | Additional Information |
|------|-------------|----------|------------------------|
| Ampere | A | electric current | - |
| Ampere-hour | Ah | electric energy | - |
| Ampere-turn | At | magnetic-field intensity | Is a CGS unit - The Oesterd is the preferred. |
| Bel | B | Audio power level | - |
| Coulomb | C | Electric Charge | - |
| Cycle per second | c/s | Frequency | Not used - the Hertz (Hz) is accepted for this unit |
| Decibel | dB | Audio power level | |
| Decibel referred to one milliwatt | dBm | | |
| Farad | F | Capacitance | |
| Gauss | G | Magnetic induction | This is a electromagnetic CGS unit. SI adopted the Tesla (T) |
| Gilbert | Gb | Magnetomotive force | This is a CGS unit. The SI unit is the ampere/turn or ampere. |
| Henry | H | Inductance | - |
| Hertz | Hz | Frequency | - |
| Horsepower | hp | Power | It isn't a SI unit but is used in many practical applications. The unit used in SI is the Watt. |
| Maxwell | Mx | Magnetic flux | This is a CGS electromagnetic unit. SI adopted the Weber |
| Mho | mho | Electric Conductance | The IEC adopted the Siemens (S) as unit of condutance |

(continued next page)

| Oesterd | Oe | Magnetic Field Intensity | This is a electromagnetic CGS unit of magnetic field strength. SI adopted the amperes per meter (A/m) |
| Ohm | $\Omega$ | Electric Resistance | |
| Revolutions per minute | r/m or rpm | | rpm is used but not recommended |
| Second | s | Time | |
| Siemens | S | Electric Conductance | mho can be found as conductance unit in old publications |
| Tesla | T | Flux Density | The Weber/$m^2$ is preferred as flux density unit. |
| Volt | V | Electric Potential | |
| Watt | W | Electric Power | |
| Watt-hour | Wh | Electric energy | |
| Weber | Wb | Magnetic Flux | 1 Wb = 1 V.s |

## TABLE 2

**Metric prefixes used in the SI to express multiples and submultiples of the units above (and others):**

| Prefix | Symbol | Value (multiply by) |
|--------|--------|---------------------|
| yotta | Y | $10^{24}$ |
| zetta | Z | $10^{21}$ |
| exa | E | $10^{18}$ |
| peta | P | $10^{15}$ |
| tera | T | $10^{12}$ |
| giga | G | $10^{9}$ |
| | (continued next page) | |

| Prefix | Symbol | Value (multiply by) |
|--------|--------|---------------------|
| mega   | M      | $10^6$              |
| kilo   | k      | $10^3$              |
| hecto  | h      | $10^2$              |
| deca   | da     | $10$                |
| deci   | d      | $10^{-1}$           |
| centi  | c      | $10^{-2}$           |
| milli  | m      | $10^{-3}$           |
| micro  | $\mu$  | $10^{-6}$           |
| nano   | n      | $10^{-9}$           |
| pico   | p      | $10^{-12}$          |
| femto  | f      | $10^{-15}$          |
| atto   | a      | $10^{-18}$          |
| zepto  | z      | $10^{-21}$          |
| yocto  | y      | $10^{-24}$          |

## TABLE 3

## Conversion Table and Applied Electric Units:

| When Converting | Into | Multiply By: |
|-----------------|------|--------------|
| amperes/sq. centimeters | amperes/sq. inches | 6,452 |
| amperes/sq. inches | amperes/sq. centimeters | 0.1550 |
| amperes-hours | coulombs | 3,600 |
| ampere-hours | faradys | 0.03731 |

(continued next page)

| | | |
|---|---|---|
| amperes-turns | gilberts | 1,257 |
| ampere-turns/cm | ampere-turns/in | 2.540 |
| ampere-turns/in | ampere-turns/cm | 0.3937 |
| coulombs | faradays | $1.036 \times 10^{-5}$ |
| coulombs | statcoulombs | $2.998 \times 10^{9}$ |
| coulombs/sq cm | coulombs/sq. in | 6.452 |
| coulombs/sq in | coulombs/sq. cm | 0.1550 |
| gausses | lines/sq. in | 6.452 |
| gausses | weber/sq. cm | $10^{-8}$ |
| gausses | weber/sq. in | $6.452 \times 10^{-8}$ |
| gilberts | ampere-turn | 0.7958 |
| gilberts/cm | amp-turns/cm | 0.7958 |
| gilberts/cm | amp-turns/in | 2.021 |
| kilowatts | Btu/min | 56.92 |
| kilowatts | horsepower | 1.341 |
| kilowatt-hours | joules | $3.6 \times 10^{6}$ |
| kilowatt-hours | Btu | 3.,413 |
| kilowatts-hour | gram-calories | 859,850 |
| ohm (international) | ohm (absolute) | 1.0005 |
| volt/in | volt/cm | 0.3937 |
| webers | maxwells | $10^{8}$ |
| webers/sq in | gausses | $1.550 \times 10^{7}$ |
| weber/sq meter | gausses | $10^{-4}$ |

## TABLE 4

## Conversion Table for Some Non-electric Important Units:

| When Converting | Into | Multiply By |
|---|---|---|
| Btu | Watts-hour | 0.293 |
| Btu | Gram-calories | 252 |
| Btu | Joules | 1,055 |
| Btu | Ft.Lbs | 778 |
| Circular mils | Square centimeters | $5.067 \times 10^{-6}$ |
| Circular mils | Square mils | 0.7854 |
| Circular mils | Square inches | $7.854 \times 10^{-7}$ |
| Cubic centimeters | Cubic inches | 0.06102 |
| Cubic inches | Cubic centimeters | 16.39 |
| Days | Seconds | 86,400 |
| Ergs | Joules | $10^{-6}$ |
| Fathoms | Feets | 6.0 |
| Ft.lbs | BTU | $1.285 \times 10^{-3}$ |
| Ft.lbs | Joules | 1.356 |
| Feet | Centimeters | 30.48 |
| Feet | Millimeters | 304.8 |
| Gallons | Cubic centimeters | 3,785.0 |
| Gallons | Cubic feet | 0.1337 |
| Gram-calories | Joules | 4.186 |
| Hand | Centimeters | 10.16 |
| Hectares | Acres | 2.471 |
| Hectares | Square feet | $1.076 \times 10^{5}$ |
| Joules | Gram-calories | 0.2388 |

(continued next page)

| | | |
|---|---|---|
| Joules | Ft.lbs | 0.7375 |
| Joules | BTUs | $9.47 \times 10^{-4}$ |
| Joules | Ergs | $10^{10}$ |
| Kilometers | Miles | 0.6214 |
| Kilometers | Yards | 1,094 |
| Kilometers | Feet | 3,281 |
| Kgs | Lbs(Avdp) | 2.2046 |
| Lbs(avdp) | Kgs | 0.4536 |
| Lbs(avdp) | Oz(avdp) | 16 |
| Lbs(Troy) | Oz(Troy) | 12 |
| Knots | Kilometers/hour | 1.8532 |
| Liters | Cubic inches | 61.02 |
| Liters | Gallons (US) | 0.2642 |
| Lumen | Watt | 0.001496 |
| Lux | Foot-candle | 0.0929 |
| Meters | Feet | 3.281 |
| Meters | Yards | 1.094 |
| Meters | Inches | 39.37 |
| Microns | Meters | $10^{-6}$ |
| Miles (statute) | Meters | 1,609 |
| Miles (nautic) | Meters | 1.853 |
| Mils | Centimeters | $2.540 \times 10^{-3}$ |
| Mils | Inches | $10^{-3}$ |
| Nepers | Decibels | 8.686 |
| Newtons | Dynes | $10^{5}$ |
| Ounces | Grams | 28.349527 |
| Ounces | Pounds | 0.0625 |
| Oz(avdp) | Lbs(avdp) | 0.0625 |
| Oz(Troy) | Lbs(Troy) | 0.0833 |

(continued next page)

| When Converting | Into | Multiply By |
|---|---|---|
| Parsecs | Miles | $19 \times 10^{12}$ |
| Parsecs | Kilometers | $3.084 \times 10^{13}$ |
| Pounds | Kilograms | 0.4536 |
| Pounds | Ounces | 16.0 |
| RPM | Degrees/Second | 6.0 |
| RPM | Radians/Second | 0.1047 |
| Radians/Second | RPM | 0.1592 |
| Square centimeters | Circular mils | $1.973 \times 10^{5}$ |
| Square inches | Circular mils | $1.273 \times 10^{6}$ |
| Square inches | Square mils | $10^{6}$ |
| Square millimeters | Circular mils | 1.973 |
| Square millimeters | Square feet | $1.076 \times 10^{-3}$ |
| Square mils | Circular mils | 1.273 |
| Square mils | Square centimeters | $6.452 \times 10^{-6}$ |
| Yards | Centimeters | 91.44 |
| Yards | Miles (stat.) | $5.682 \times 10^{-4}$ |
| Watt-hour | BTU | 3.416 |

NOTE: All the calculations given as examples in this book will use, as preferred units, the ones in SI. But, using the conversion tables the reader can easily covert them if desired.

## TABLE 5

## Numerical Latin/Greek Prefixes

| Number | Greek Prefix | Latin Prefix |
|---|---|---|
| 1/2 | hemi | semi |
| 1 | mono/mon | uni |
| 2 | di | bi/duo |
| (continued next page) | | |

| | | |
|---|---|---|
| 3 | tri | tri/ter |
| 4 | tetra/tetr | quadri/quadr |
| 5 | oenta/pent | quinque/quinqu |
| 6 | hexa/hex | sexi/sex |
| 7 | hepta/hept | septi/sept |
| 8 | octa/octo | octo |
| 9 | enne/ennea | nona/novem |
| 10 | deca/dec | decem |
| 11 | hendeca/hendec | undeca/undec |
| 12 | dodeca/dodec | duodec |
| 13 | trideca/tridec | tredec |
| 14 | tetradeca/tetradec | quatrodec |
| 15 | pentadec/pentadec | quindec |
| 16 | hexadec/hexadec | sextodec |
| 17 | heptadec/heptadec | septendec |
| 18 | octadeca/octadec | octodec |
| 19 | nonadeca/nonadec | novendec |
| 20 | eicosa/eicos | - |
| 21 | heneicosa/heneicos | - |
| 22 | docosa/docos | - |
| 23 | tricosa/tricos | - |
| 24 | tetracosa/tetracos | - |
| 25 | pentacosa/pentacos | - |
| 26 | hexacosa/hexacos | - |
| 27 | heptacosa/heptacos | - |
| 28 | octacosa/octacos | - |
| 29 | nocacosa/nonacos | - |
| 30 | triaconta/triacont | tringti |
| 31 | hentriaconta/hentriacont | - |
| 32 | dotriaconta/dotriacont | - |
| 40 | tetraconta/tetracont | quadragin |
| 50 | pentaconta/pentacont | quinquigin |

# 2. RESISTANCE OF A CONDUCTOR

The resistance of an homogeneous conductor depends on its length, cross-sectional area and the nature of the material The next formula is used to calculate the resistance from those quantities:

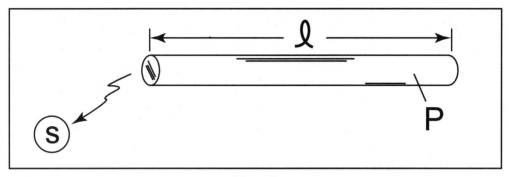

*Figure 1*

## Formula 2.1

$$R = \rho \, \frac{l}{S}$$

Where:

R = conductor's resistance in ohms ( $\Omega$ )

$\rho$ = resistivity or specific resistance of the used material  (see table below)

l = length in meters (m)

S = cross-sectional area in square milimeters

## DERIVATED FORMULAS

## Formula 2.2

$$l = R \, \frac{S}{\rho}$$

## Formula 2.3

$$S = \rho \, \frac{l}{R}$$

**Application Example:**

What is the resistance of a length of homogeneous copper wire with a constant cross sectional area of 0.5 mm² ? The piece of wire has a length of 1m.

Data:

$l = 1$ m

$\rho = 0.016$ (see table)

$S = 0.5$ mm²

$R = ?$

Applying Formula 2.1:

$$R = 0.016 \times \frac{1}{0.5} = 0.032 \ \Omega$$

## TABLE 6

### Specific Resistance Of Some Materials

In ohms by meter of length and square milimeters of cross-sectional area.

| Material | $\rho$ |
|----------|--------|
| Aluminium | 0.0292 |
| Antimony | 0.417 |
| Bismuth | 1.17 |
| (continued next page) | |

| Material | $\rho$ |
|----------|--------|
| Bronze | 0.067 |
| Cadmium | 0.076 |
| Copper (pure) | 0.0162 |
| Copper (hard) | 0.0178 |
| Constantan | 0.5 |
| Graphite | 13 |
| Gold | 0.024 |
| Iron (pure) | 0.096 |
| Lead | 0.22 |
| Mercury | 0.96 |
| Nickel | 0.087 |
| Platinum | 0.106 |
| Silver | 0.0158 |
| Stain | 0.115 |
| Tin | 0.067 |
| Tungsten | 0.055 |
| Zinc | 0.056 |

# 3. CONDUCTANCE (I)

Conductance is defined as the inverse of resistance. The first unit assigned to this quantity was the mho (ohm inverted), but now we use the Siemens (S) as unit of conductance.

To calculate the conductance from the resistance we use:

## Formula 3.1

$$G = \frac{1}{R}$$

Where:

G is the conductance in Siemens (S)

R is the resistance in ohms ( $\Omega$ )

## DERIVATED FORMULAS

### Formula 3.2

$$R = \frac{1}{G}$$

**Application Example:**

What is the conductance of a 25-ohm resistor?

Data:

R = 25 $\Omega$

G = ?

Applying Formula 3.1:

$$G = \frac{1}{25} = 0.04 \text{ S}$$

# 4. CONDUCTANCE OF A LENGTH OF WIRE

This formula is used to calculate the conductance of a wire as a function of its length, cross-sectional area and nature of the material.

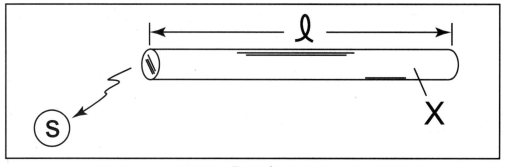

*Figure 2*

## Formula 4.1

$$G = \chi \, \frac{S}{l}$$

Where:

G is the conductance in siemens (S)

$\chi$ is the specific conductance or conductivity of the material

S is the cross-sectional area in square milimeters (mm²)

l is the length in meters (m)

## DERIVATED FORMULAS

## Formula 4.2

$$S = G \, \frac{l}{\chi}$$

## Formula 4.3

$$l = \chi \, \frac{S}{G}$$

## Application Example:

What is the conductance of a piece of aluminium wire 2 meters long and with a cross-sectional area of 0.1 mm²?

Data:

$\chi$ = 34.4 (specific conductance for aluminium - see table)

l = 2 m

S = 0.1 mm²

G = ?

Applying Formula 4.1:

$$G = 34.4 \text{ x } \frac{0.1}{2} = 1.72 \text{ S}$$

## Additional Formula:

To find the cross-sectional area when having the diameter of a circular-section wire use the next formula:

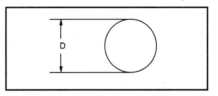

## Formula 4.A

$$S = \pi \text{ x } \frac{D^2}{4}$$

Where:

S is the sectional area in square milimeters (mm²)

$\pi$ is the constant 3.14

D is the wire's diameter in milimeters (mm)

## TABLE 7

### Conductance of Some Common Materials

In Siemens by square milimeter of cross-sectional area and meter of length.

| Material | $\chi$ |
|----------|--------|
| Aluminum | 34.2 |
| Antimony | 2.5 |
| Bronze | 14.9 |
| Bismuth | 0.85 |
| Cadmium | 13 |
| Cobalt | 10.4 |
| Copper (pure) | 61.7 |
| Copper (hard) | 56.1 |
| Constantan | 2 |
| Gold | 43.5 |
| Graphite | 0.07 |
| Iron (pure) | 10.2 |
| Lead | 4.8 |
| Mercury | 1.044 |
| Nicrome | 0.909 |
| Nickel | 10.41 |
| Platinum | 9.09 |
| Silver | 62.5 |
| Stainless | 8.6 |
| Tungsten | 18.18 |
| Zinc | 17.8 |

NOTE: In real applications the values can change slightly from the ones indicated in the table according to the degree of purity or the composition of the material.

# 5. Thermal Influence in the Resistance of a Cable

When the temperature changes the resistivity of the materials in electric wires and cables also changes. The alterations can be calculated by the formulas given below and the constants of the used materials as given in Table 3.

## Formula 4.1

$$\mathbf{R}_{t1} = \mathbf{R}_{t2} \left[1 + \alpha(t1 - to)\right]$$

Where:

$R_{t1}$ is the final resistance in ohms ($\Omega$)

$R_{t0}$ is the initial resistance in ohms ($\Omega$)

t1 is the final temperature (in degrees Celsius or Kelvin)

t0 is the initial temperature (in degrees Celsius or Kelvin)

$\alpha$ is the temperature coefficient of the used material (see table)

## TABLE 8

## Temperature Coefficient of Resistivity of Some Materials at 20°C

| Material | Coefficient ($\alpha$) |
|---|---|
| Aluminum | 0.00002 |
| Bronze | 0.002 |
| Copper | 0.00382 |
| Constantan | 0.00001 |
| Tin | 0.002 |
| Nicrome | 0.00013 |
| Mercury | 0.00089 |
| (continued next page) | |

| Material | Coefficient ($\alpha$) |
|----------|------------------------|
| Stainless | 0.0042 |
| Gold | 0.0034 |
| Silver | 0.0038 |
| Platinum | 0.0025 |
| Zinc | 0.0038 |
| Nickel | 0.0047 |

NOTE:  According the composition or degree of purity of the considered material the real values can present small differences when compared to values shown in the table.

## TABLE 9
## Standard Annealed Copper Wire (AWG & B&S)

| AWG | Diameter | Cross-section (mm²) | Resistance (ohms/km) |
|-----|----------|---------------------|----------------------|
| 0000 | 11.86 | 107.2 | 0.158 |
| 000 | 10.40 | 85.3 | 0.197 |
| 00 | 9.226 | 67.43 | 0.252 |
| 0 | 8.252 | 53.48 | 0.317 |
| 1 | 7.348 | 42.41 | 0.40 |
| 2 | 6.544 | 33.63 | 0.50 |
| 3 | 5.827 | 26.67 | 0.63 |
| 4 | 5.189 | 21.15 | 0.80 |
| 5 | 4.621 | 16.77 | 1.01 |
| 6 | 4.115 | 13.30 | 1.27 |
| 7 | 3.665 | 10.55 | 1.70 |
| 8 | 3.264 | 8.36 | 2.03 |
| 9 | 2.906 | 6.63 | 2.56 |
| 10 | 2.588 | 5.26 | 3.23 |
| 11 | 2.305 | 4.17 | 4.07 |
| 12 | 2.053 | 3.31 | 5.13 |

| | | | |
|----|--------|--------|-------|
| 13 | 1.828  | 2.63   | 6.49  |
| 14 | 1.628  | 2.08   | 8.17  |
| 15 | 1.450  | 1.65   | 10.3  |
| 16 | 1.291  | 1.31   | 12.9  |
| 17 | 1.150  | 1.04   | 16.34 |
| 18 | 1.024  | 0.82   | 20.73 |
| 19 | 0.9116 | 0.65   | 26.15 |
| 20 | 0.8118 | 0.52   | 32.69 |
| 21 | 0.7230 | 0.41   | 41.46 |
| 22 | 0.6438 | 0.33   | 51.5  |
| 23 | 0.5733 | 0.26   | 56.4  |
| 24 | 0.5106 | 0.20   | 85.0  |
| 25 | 0.4547 | 0.16   | 106.2 |
| 26 | 0.4049 | 0.13   | 130.7 |
| 27 | 0.3606 | 0.10   | 170.0 |
| 28 | 0.3211 | 0.08   | 212.5 |
| 29 | 0.2859 | 0.064  | 265.6 |
| 30 | 0.2546 | 0.051  | 333.3 |
| 31 | 0.2268 | 0.040  | 425.0 |
| 32 | 0.2019 | 0.032  | 531.2 |
| 33 | 0.1798 | 0.0254 | 669.3 |
| 34 | 0.1601 | 0.0201 | 845.8 |
| 35 | 0.1426 | 0.0159 | 1 069 |
| 36 | 0.1270 | 0.0127 | 1 339 |
| 37 | 0.1131 | 0.0100 | 1 700 |
| 38 | 0.1007 | 0.0079 | 2 152 |
| 39 | 0.0897 | 0.0063 | 2 669 |
| 40 | 0.0799 | 0.0050 | 3 400 |
| 41 | 0.0711 | 0.0040 | 4 250 |
| 42 | 0.0633 | 0.0032 | 5 312 |
| 43 | 0.0564 | 0.0025 | 6 800 |
| 44 | 0.0503 | 0.0020 | 8 500 |

# 5. OHM'S LAW

The voltage drop accross a resistor is equal the product of the current passing through it by its resistance.

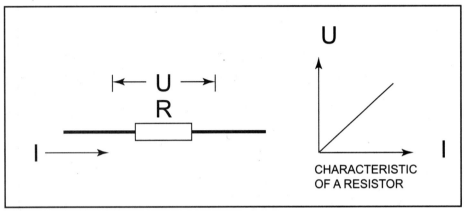

*Figure 3*

## Formula 5.1

$$U = R \times I$$

Where:

U is the voltage in Volts (V)

I is the current flowing through the resistor in amperes (A)

R is the resistance in ohms ($\Omega$)

## Derivated Formulas:

## Formula 5.2

$$I = \frac{U}{R}$$

**Formula 5.3**

$$R = \frac{U}{I}$$

**Application Example:**

Calculate the current flowing through a 100-ohm resistor connected to a 20V power supply.

Data:

>   R = 100 ohms
>
>   V = 20 Volts
>
>   I = ?

Applying Formula 5.2:

$$I = \frac{20}{100} = 0.2 \text{ A or } 200 \text{ mA}$$

# 6. POWER

The amount of power consumed by a device or drained from a generator is proportional to the product of the amount of current flowing through the device by the applied voltage.

**Formula 6.1**

$$P = U \times I$$

Where:

>   P is the power in Watts (W)
>
>   U is the voltage in volts (V)
>
>   I is the current in amperes (A)

## DERIVATED FORMULAS

### Formula 6.2

$$U = \frac{P}{I}$$

### Formula 6.3

$$I = \frac{P}{U}$$

**Application Example:**

An incandescent lamp draws 200 mA when plugged to a 6V power supply. What is the power consumed by this lamp?

Data:

> I = 200 mA or 0.2 A
>
> V = 6 VDC
>
> P = ?

Applying Formula 6.1:

> P = 0.2 x 6 = 1.2 W or Joules/second (J/s)

# 7. JOULE'S LAW

The amount of electric power transformed in heat in a conductor (pure resistance) is equal to the product of the resistance of the conductor by the square of the amount of current flowing through it.

## Formula 7.1

$$P = R \times I^2$$

Where:

P is the amount of power converted to heat in watts (W)

R is the resistance of the conductor in ohms ($\Omega$)

I is the current in amperes (A)

## Derivated Formulas:

## Formula 7.2

$$R = \frac{P}{I2}$$

## Formula 7.3

$$I = \sqrt{\frac{P}{R}}$$

## Formula 7.4

$$P = \frac{U^2}{R}$$

## Formula 7.5

$$U = \sqrt{PxR}$$

**Formula 7.6**

$$R = \frac{U^2}{P}$$

**Application Example:**

Calculate the amount of power converted into heat by a 100-ohm resistor with a 2A current flowing through it.

Data:

R = 100 ohms

I = 2 ampere

P = ?

Using Formula 7.1:

P = 100 x $2^2$ = 100 x 4 = 400 watts

# 8. ELECTRIC ENERGY

The energy consumed by a (or furnished to) device is equal to the product of the power by the time interval the device is on.

**Formula 8.1**

$$W = P \times t$$

Where:

W is the amount of consumed energy in watts/second (Ws) or Joules (J)

P is the device's power in watts (W) or joules/second (J/s)

t is the time the device is ON in seconds (s)

## Derivated Formulas:

## Formula 8.2

$$P = \frac{W}{t}$$

## Formula 8.3

$$t = \frac{W}{P}$$

**Application Example:**

How much energy is consumed by a 100W incandescent lamp in 10 minutes?

Data:

P = 100W

t = 10 minutes = 600 seconds

W = ?

Using Formula 8.1:

W = 100 x 600 = 60 000 Joules

NOTE: The consumed (or furnished) energy can also be calculated in kilowatts-hour (kWh). For the application example the calculation is made as follows:

Data :

P = 100 W = 0.1 kW

t = 10 minutes = 1/6 hour = 0.133 hour

W = ?

Using again Formula 8.1:

$$W = 0.1 \times 0.133 = 0.0133 \text{ kWh}$$

# 9. ELECTROLYSIS (Faraday's Law)

The mass of any substance liberated at the electrode in electrolysis is proportional to the mass of a given substance liberated when unit a quantity of charge passes through the electrolyte.

## Formula 9.1

$$G = g.l.t$$

Where:

G is the mass of deposited substance in milligrams (mg)

g is the chemical equivalent of the involved substance in mg/As

I is the amount of current flowing between electrodes in amperes (A)

t is the electrolysis time in seconds (s)

## TABLE 10
## Electrochemical Equivalents of Some Substances

| Íon | g (mg/As) |
|-----|-----------|
| H+ | 1.0104 |
| O− | 0.0829 |
| Al+++ | 0.936 |
| OH- | 0.1762 |
| Fe+++ | 0.1930 |
| Ca++ | 0.2077 |
| (continued next page) | |

| | |
|---|---|
| Na+ | 0.2388 |
| Fe++ | 0.2895 |
| $CO_3$— | 0.3108 |
| Cu++ | 0.3297 |
| Zn++ | 0.3387 |
| Cl- | 0.3672 |
| $SO_4$— | 0.4975 |
| Cu+ | 0.6590 |
| Ag+ | 1.118 |

## Derivated Formulas:

## Formula 9.2

$$I = \frac{G}{g.t}$$

## Formula 9.3

$$t = \frac{G}{g.I}$$

## Application Example:

Calculate the mass of silver (Ag) deposited in an electrolysis process when a 2A current flows for 10 minutes.

Data:

g = 1.118 mg/As

I = 2A

t = 10 minutes = 600 seconds

Applying formula 9.1:

$$G = 1.118 \times 2 \times 600 = 1,341 \text{ mg or } 1.341 \text{ g}$$

# 10. RESISTORS IN SERIES

The equivalent resisistance of resistors wired in parallel is equal to the sum of the resistances of each associated resistor.

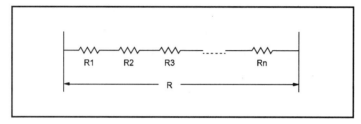

*Figure 4*

### Formula 10.1

$$R = R_1 + R_2 + R_3 + \ldots\ldots\ldots + R_n$$

Where:

R is the equivalent resistance in ohms ($\Omega$)

R1 to Rn are the associted resistances in ohms ($\Omega$)

### Derivated Formulas:

Particular case:

All the associated resistances are equal (R1=R2=R3....=Rn).

### Formula 10.2

$$R = n \times R1$$

Where:

> R is the equivalent resistance in ohms ($\Omega$)
>
> R1 is the value of the associated resistances in ohms ($\Omega$)
>
> n is the number of associated resistances

**Important Properties:**

- The greatest resistance converts the greatest amount of power in heat.
- The amount of current flowing by all resistors is the same.
- The equivalent resistance is greater than the the greatest associated resistance.

**Application Example:**

What is the equivalent resistance of an association where resistors of 20, 30 and 100 ohms are wired in series? Which resistor dissipates more heat?

Data:

> R1 = 20 ohms
>
> R2 = 30 ohms
>
> R3 = 100 ohms
>
> R - ?

Using Formula 10.1:

$$R = 20 + 30 + 100 = 150 \text{ ohms}$$

R3 (100 ohms) converts more power in heat. Use Ohm's Law and Joule's Law to calculate the amount of power converted in heat when the association is submitted to a certain voltage.

# 11. RESISTORS IN PARALLEL

The equivalent conductance of a parallel association of resistors is equal to the sum of the conductances of the associated resistors. As conductance is defined as the inverse of resistance, we can write two formulas to calculate resistors wired in parallel:

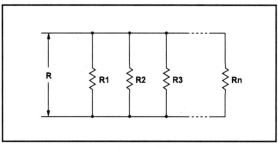

*Figure 5*

## Formula 11.1

$$G = G1 + G2 + G3 + ..... + Gn$$

Where:

G is the equivalent conductance of the association in siemens (S)

G1, G2, G3.....Gn are the conductances of the associated resistors in Siemens (S)

## Derivated Formulas:

## Formula 11.2

$$\frac{1}{R} = \frac{1}{R1} + \frac{1}{R2} + \frac{1}{R3} + ............ + \frac{1}{Rn}$$

Where:

R is the equivalent resistance in ohms ($\Omega$)

R1, R2, R3....Rn are the associated resistances in ohms ($\Omega$)

**Particular Case:**

**Equal Resistances in Parallel**

If n equal resistors are associated in parallel the equivalent resistance can be calculated by the next formula:

## Formula 11.3

$$R = \frac{R1}{R}$$

Where:

R is the equivalent resistance in ohms

R1 is the value of all the associated resistors (all equal)

**Important Properties:**

- The equivalent resistance is lower than the lowest associated resistance.
- The lowest resistance converts more power in heat.
- The highest current flows through the lowest resistance.

**Application Example:**

Calculate the equivalent resistance of 20-ohm and 30-ohm resistors wired in parallel.

Data:

R1 = 20 ohms

R2 = 30 ohms

R = ?

Applying Formula 11.2:

1/R = 1/20 + 1/30

Converting the terms in the second member of the equality to the same denominator:

1/R = 3/60 + 2/60

1/R = 5/60

Inverting both members in the equality:

R = 60/5 = 12 ohms

# Table 11

## Color Code for Resistors

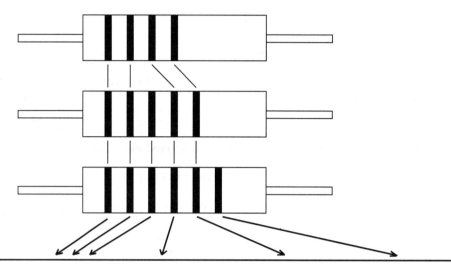

| Color | Significant Figures | Multiplier | Tolerance | Temperature Coefficient |
|-------|---------------------|------------|-----------|-------------------------|
| ppm/°C | | | | |
| black | 0 | 1 | - | - |
| Brown | 1 | 10 | 1% | 100 |
| Red | 2 | 100 | 2% | 50 |
| Orange | 3 | 1 000 | - | 15 |
| Yellow | 4 | 10 000 | - | 25 |
| Green | 5 | 100 000 | 0.5% | - |
| Blue | 6 | 1 000 000 | 0.25% | 10 |
| Violet | 7 | 10 000 000 | 0.1% | 5 |
| Gray | 8 | 100 000 000 | 0.05% | - |
| White | 9 | 1 000 000 000 | - | 1 |
| Gold | - | 0.1 | 5% | - |
| Silver | - | 0.01 | 10% | - |

# 12. VOLTAGE DIVIDER

The next formulas are used to calculate the voltage accross resistors in a series association or voltage divider as shown below.

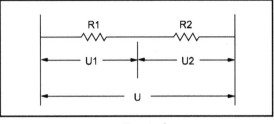

*Figure 6*

# Formulas 12.1

$$U1 = U \times \frac{R1}{R1+R2} \qquad U2 = U \times \frac{R2}{R1+R2}$$

Where:

U is the voltage applied to the divider in volts (V)

U1 and U2 are the voltages accross the resistors R1 and R2 (V)

R1 and R2 are the resistances in ohms ($\Omega$)

**Application Example:**

Two resistors (R1 = 20 ohms and R2 = 30 ohms) are wired as a voltage divider and plugged to a 10-volt power supply. Calculate the voltage accross R1.

Data:

R1 = 20 ohms

R2 = 30 ohms

U = 10 Volts

U1 = ?

Applying Formula 12.1:

$$U1 = 10 \text{ x } \frac{20}{20+30} = 10 \text{ x } \frac{20}{50} = 4 \text{ volts}$$

# 13. LOADED VOLTAGE DIVIDER

When a device drains current from a voltage divider loading it, the voltage must be calculated by the next formula:

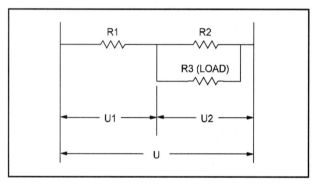

*Figure 7*

## Formula 13.1

$$U2 = U \text{ x } \frac{R2xR3}{R1xR2 + R2xR3 + R1xR3}$$

Where:

U2 is the voltage accross the loaded resistor (R2) (V)

U is the voltage applied to the divider (V)

R1 and R2 are the resistors in the voltage divider in ohms ($\Omega$)

R3 is the load resistance in ohms ($\Omega$)

# 14. FIRST KIRCHHOFF'S LAW

In a considered point of a circuit the algebraic sum of the currents flowing into a junction (or branch point) is zero. In the circuit below the currents flowing to the point are positive and the currents flowing from the point are negative.

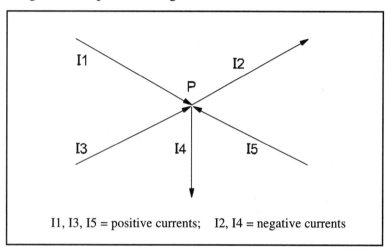

I1, I3, I5 = positive currents;    I2, I4 = negative currents

*Figure 8*

## Formula 14.1

$$I1 + I2 + I3 + I4 + I5 = 0$$

Where:

I1, I2, I3, I4, I5 are the currents flowing to the junction P in amperes (A)

P is the junction point or branch point

NOTE: For the circuit given as example we can write:

I1 - I2 + I3 - I4 + I5 = 0

# 15. SECOND KIRCHHOFF'S LAW

The algebraic sum of the products of the currents by the respective resistance around a closed loop is equal to the algebraic sum of the emf's in the loop.

NOTE: An emf is considered positive if the arbitrary direction around the loop coincides with the direction of the emf of the current source. *Figure 9* gives the example to the formula application:

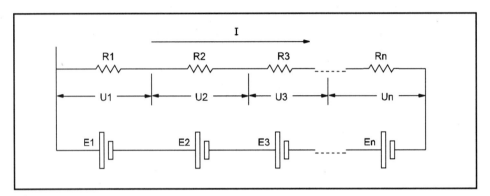

*Figure 9*

**Formula 15.1**

$$E1 + E2 + E3 +.....+ En = U1 + U2 + U3 +......+ Un$$

Where:

E1, E2, E3.... En are the emf in volts (V)

U1, U2, U3...Un are the voltage falls accross the resistors R1, R2, R3....Rn in volts.

Considering that U1 = R1 x I, U2 = R2 x I, U3 = R3 x I .......Un = Rn x I, we can write the next formula:

**Derivated Formula:**

**Formula 15.2**

$$E1 + E2 + E3 + .... + En = R1xI + R2xI + R3xI + ...... + RnxI$$

---

Where:

E1, E2, E3.... En are the emf in volts (V)

R1, R2, R3....Rn the resistances of the resistors in the loop in ohms ( $\Omega$ )

I is the current flowing through the circuit in amperes (A)

# 16. CAPACITANCE

The capacitance of a capacitor is defined as the ratio of the charge in one of the plates to the potential difference between the plates. The next formulas are used to make calculations involving capacitances.

## Formula 16.1

$$C = \frac{Q}{U}$$

Where:

C is the capacitance in Farads (F)

Q is the amount of charge in one plate in coulombs (C)

U is voltage between plates in volts (V)

NOTE: If the capacitance is expressed in $\mu$ F, the charge will be found in $\mu$ C or, if the capacitance is expressed in pF the charge will be found in pC (keeping the voltage in volts).

## Derivated Formulas:

## Formula 16.2

$$Q = C \times U$$

**Formula 16.3**

$$U = \frac{Q}{C}$$

**Application Example:**

Calculate the amount of charge stored in a 100 $\mu$F capacitor when the potential difference between plates is 100 volts.

Data:

$C = 100$ uF $= 100 \times 10^{-6}$ C

$U = 100$V

$Q = ?$

Applying Formula 16.2:

$Q = 100 \times 10^{-6} \times 100 = 10^3 \times 10^{-6} = 10^{-3}$ mC or $1\ 000\ \mu$C

# 17. PLANAR CAPACITOR

A planar capacitor is formed by two parallel plates and between them is placed a isolator called dieletric. The capacitance depends upon the size of the plates, the distance between them and the nature of the material used as dieletric.

**Formula 17.1**

$$C = 0.08859 \times \varepsilon \times \frac{s(n-1)}{d}$$

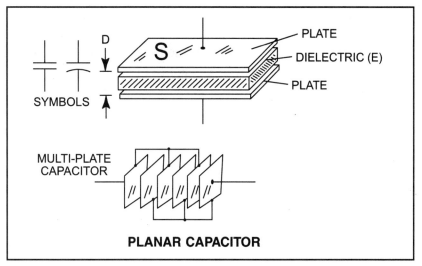

*Figure 10*

Where:

C is the capacitance in picofarads (pF)

$\varepsilon$ is the dieletric constant (see table)

S is the active surface area of the smaller plate (if the capacitor uses different size plates in square centimeters (cm²)

n is the number of plates

d is the distance between plates in centimeters (cm)

## Derivated Formulas:

### Formula 17.2

$$d = 0.08859 \times \varepsilon \times \frac{s(n-1)}{d}$$

### Formula 17.3

$$s = \frac{dxC}{0.08858x\varepsilon x(n-1)}$$

**Application Example:**

What is the capacitance of a planar capacitor formed by two equal plates with an effective area of 100 square centimeters, separated by a distance of 2 cm and using a piece of glass as dieletric?

Data:

$s = 100 \text{ cm}^2$

$d = 2 \text{ cm}$

$n = 2$

$\varepsilon = 10$ (see table)

$C = ?$

Using Formula 17.1:

$$C = 0.08859 \text{ x } 10 \text{ x } \frac{100x(2-1)}{2} = 0.08859 \text{ x } 10 \text{ x } 50$$

$$C = 44.295 \text{ pF}$$

## TABLE 12
## Dielectric Constant for Some Materials

The numbers shown in this table in many cases are averages, depending on the manufacturing processes or the composition, and in particular cases they can assume a wide range of values.

| Material | Dieletric Constant ($\varepsilon$) |
| --- | --- |
| Air | 1.0 |
| Acetone | 21 to 23 |
| Bakelite | 4.5 to 7 |
| (continued next page) | |

| | |
|---|---|
| Beeswax | 2.7 to 2.9 |
| Benzene | 2.2 t 2.3 |
| Bitumen | 2.5 to 3.3 |
| Celluloid | 3 to 4 |
| Ebonite | 2.8 to 4.5 |
| Ethyl Alcohol | 20 to 27 |
| Ethyl Ether | 4.1 to 4.8 |
| Glass | 6 to 10 |
| Glycerine | 50 to 56 |
| Kerosene | 2.0 |
| Marble | 8 to 10 |
| Mica | 2.5 to 8 |
| Parafin | 2 to 2.5 |
| Plexiglass | 3 to 3.5 |
| Polystyrene | 2.2 to 2.5 |
| Polyvinyl | 3.0 to 3.6 |
| Porcelain | 3.1 to 6.5 |
| Rosin | 2.5 to 3.5 |
| Rubber (soft) | 2.6 to 3.0 |
| Silk (natural) | 4.5 |
| Vacuum | 0.99 |

NOTE: Liquid substances also show changes in the dieletric constant with the temperature.

# 18. BREAKDOWN VOLTAGE IN A CAPACITOR

If the voltage between the plates of a capacitor rises, passing over a certain value, breakdown occurs. The dieletric turns into a conductor and the capacitor is damaged by the produced spark. The maximum voltage that can be applied to a capacitor without causing rupture of the dieletric is called *breakdown voltage* and is calculated as follows:

## Formula 18.1

$$U_b = D_s \times d$$

Where:

$U_b$ is the breakdown voltage in kilovolts (kV)

$D_s$ is the dieletric strength in kV/mm (see table)

$d$ is the distance between the plates in milimeters (mm)

## Derivated Formulas:

## Formula 18.2

$$D_s = \frac{Ub}{d}$$

## Formula 18.3

$$d = \frac{U_b}{Ds}$$

**Application Example:**

Calculate the highest voltage that can be applied between the plates of a capacitor formed by two plates separated by a distance of 5 mm and using a piece of glass as dieletric?

Data:

$d = 5$ mm

$Ds = 20$ kV/mm (from the table)

$Ub = ?$

Applying Formula 18.1:

$$Ub = 20 \times 5 = 100 \text{ kV}$$

## TABLE 13
### Dielectric Strength of Some Materials

According the nature of the material, origin, and also particular composition, the values shown in the table can assume a certain range of voltages. The table shows average values and is useful as starting point in noncritical projects. For critical projects consult the particular characteristics of the material.

| Material | Dieletric Strength (kV/mm) |
|---|---|
| Asbestos | 4 to 4.6 |
| Bakelite | 10 to 40 |
| Beeswax | 20 to 35 |
| Bitumen | 5 to 15 |
| Celluloid | 25 to 30 |
| Ebonite | 20 to 25 |
| Fibre board (dry) | 2 to 6 |
| Glass | 20 to 30 |
| Marble | 6 to 10 |
| Mica | 40 to 200 |
| Parafin | 8 to 15 |
| Plexiglass | 15 to 19 |
| Polystyrene | 25 to 50 |
| Polyvinyl | 40 to 50 |
| Rubber (soft) | 15 to 25 |

# 19. ENERGY STORED IN A CAPACITOR

The space between the plates of a charged capacitor is filled by an electric field where the energy is stored. The stored energy in a capacitor can be calculated by the following formulas:

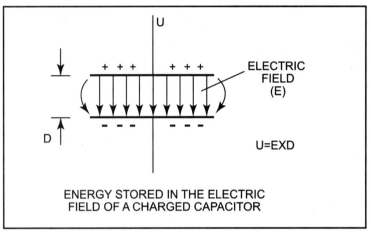

ENERGY STORED IN THE ELECTRIC
FIELD OF A CHARGED CAPACITOR

*Figure 11*

## Formula 19.1

$$W = \frac{1}{2} \times C \times U^2$$

Where:

W is the stored energy in joules (J)

C is the capacitance in farads (F)

U is the voltage between the capacitor's plates in volts (V)

## Derivated Formulas:

## Formula 19.2

$$W = \frac{1}{2} \times Q \times U$$

Where:

Q is the charge in one of the plates in coulombs (C)

W is the stored energy in joules (J)

U is the voltage between plates

## Application Example:

In a xenon flasher a 200 uF capacitor is charged with 300V from a DC converter. How much energy can be delivered by the xenon lamp in one flash (considering the total discharge when releasing the stored energy)?

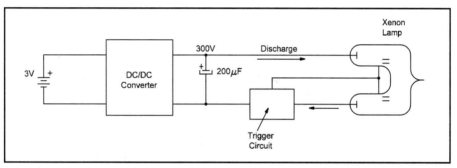

*Figure 12*

Data:

$$C = 300 \ \mu F = 300 \times 10^{-6} \ F$$

$$U = 300 \ V$$

Applying Formula 19.1:

$$W = 0.5 \times 300 \times 10^{-6} \times 300 \times 300$$

$$W = 45000 \times 10^{-6}$$

$$W = 45 \times 10^{-3}$$

$$W = 45 \ mJ$$

# 20. CAPACITORS IN PARALLEL

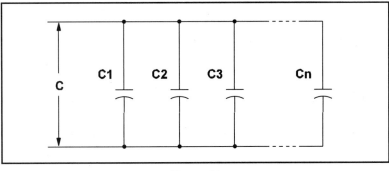

*Figure 13*

## Formula 20.1

$$C = C1 + C2 + C3 + ..... + Cn$$

Where:

C1, C2, C3....Cn are parallel asssociated capacitors in farads (F) or any submultiple

C is the equivalent capacitance in farads (F) or any submultiple

## Derivated Formulas:

## Formula 20.2

If n capacitors having the same capacitance (C1) are associated in parallel the equivalent capacitance C is calculated by the following "reduced" formula:

$$C = n \times C1$$

Important Properties:

• The equivalent capacitance is larger than the largest associated capacitor.

• All the capacitors are submitted to the same voltage.

• The largest capacitor stores the largest charge.

**Application Example:**

Calculate equivalent capacitance when wiring parallel capacitors of 20 $\mu$ F, 40 $\mu$ F and 50 $\mu$ F.

Data:

C1 = 20 $\mu$ F

C2 = 40 $\mu$ F

C3 = 50 $\mu$ F

Applying Formula 20.1

C = 20 + 40 + 50 = 110 $\mu$ F

# 21. CAPACITORS IN SERIES

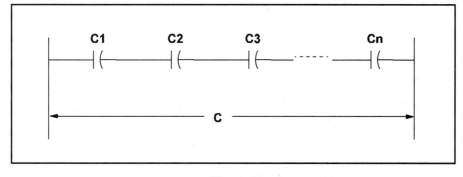

*Figure 14*

## Formula 21.1

$$\frac{1}{C} = \frac{1}{C1} + \frac{1}{C2} + \frac{1}{C3} + \text{......} + \frac{1}{Cn}$$

Where:

C1, C2, C3.....Cn  are the associated capacitors (in Farads or submultiples)

C is the equivalent capacitance in Farads or other submultiple

## Derivated Formula:

## Formula 21.2

If n equal capacitors (C1) associated in series, the equivalent capacitance is calculated by the next formula:

$$C = \frac{C1}{n}$$

## Formula 21.3

For two capacitors (C1 and C2) the formula is reduced to:

$$C = \frac{C1 \times C2}{C1 + C2}$$

Important Properties:

- All capacitors store the same charge.

- The smallest capacitor is submitted to the higher voltage.

- The equivalent capacitance is smaller than the smallest associated capacitor.

**Application Example:**

What is the equivalent capacitance when wiring a 20 pF in series with a 30 pF capacitor?

Data:

    C1 = 20 pF

    C2 = 30 pF

    C = ?

Using Formula 21.1 (you can also use Formula 21.3):

$$\frac{1}{C} = \frac{1}{20} + \frac{1}{30} = \frac{3+2}{60} = \frac{5}{60}$$

$$C = \frac{60}{5} = 12 \text{ pF}$$

# 22. MAGNETIC FIELD IN A SOLENOID

The strength of a magnetic field inside an air-core cylindrical coil (solenoid) depends on the amount of current flowing through it, the number of turns, and length of the coil.

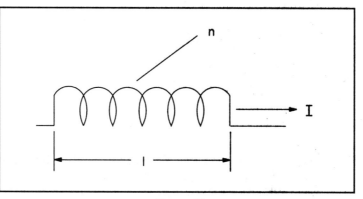

*Figure 15*

**Formula 22.1**

$$H = 1.257 \times \frac{Ixn}{L}$$

Where:

H is the magnetic field strength in oersteds (Oe)

I is the current flowing through the solenoid in amperes (A)

L is the length of the solenoid in centimeters (cm)

n is the number of turns

## Derivated Formulas:

### Formula 22.2

$$I = \frac{H x L}{1.257 x n}$$

### Formula 22.3

$$H = \frac{H x L}{1.257 x I}$$

### Formula 22.4

$$L = 1.257 \text{ x } \frac{I x n}{H}$$

**Application Example:**

A 2 ampere current flows through an air-core solenoid formed by 60 turns of wire 3 cm long. Calculate the field strength created by this solenoid.

Data:

    I = 2 A

    n = 60

    L = 3 cm

    H = ?

Applying Formula 22.1:

$$H = 1.257 \text{ x } \frac{2 x 60}{3} = 1.357 \text{ x } 40 = 50.28 \text{ Oe}$$

# 23. MAGNETIC INDUCTION INSIDE A SOLENOID

The magnetic induction inside a solenoid with a ferromagnetic core depends on the current flowing through the solenoid, the number of turns, the length and also on the permeability of the material used in the core.

The properties of ferromagnetic substances are explained by means of the Domain Theory of Magnetization.

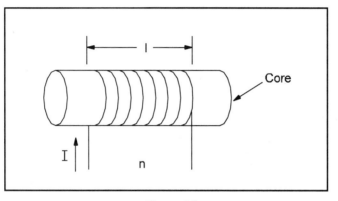

*Figure 16*

## Formula 23.1

$$H = 1.257 \times \frac{nxI}{L} \times \mu$$

Where:

H is magnetic induction in gauss (Gs)

I is the curent flow through the solenoid in amperes (A)

n is the number of turns

L is the solenoid's length in centimeters (cm)

$\mu$ is the permeability of the material used in the core

NOTE: Materials with $\mu$ larger than 1 are called *paramagnetic* and materials with $\mu$ smaller than 1 are called *diamagnetic*.

# 24. INDUCTANCE

Any change in the current in a solenoid leads to the appearance of an induced emf due to the magnetic flux of the current. This phenomenon is called *self-induction*.

The self-inductance or inductance of a air-core coil can be calculated by the following formula:

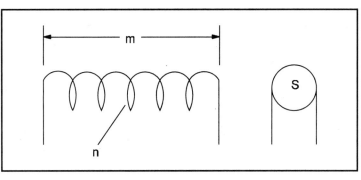

*Figure 17*

## Formula 24.1

$$L = 1.257 \times \frac{Sxn^2}{m} \times 10^{-8}$$

Where:

      L is the self-induction coefficient in henrys (H)

      n is the number of turns

      S is the cross-section area of a turn in square centimeters (cm$^2$)

      m is the solenoid's length in centimeters (cm)

NOTE: The henry (H) is the self-inductance unit and is defined as the inductance of a conductor in which a current change of 1 ampere per second induces an emf of 1 volt.

## Derivated Formulas:

### Formula 24.2

$$n = \sqrt{\frac{Lxmx10^8}{1.257xS}}$$

### Formula 24.3

$$S = 10^8 \times \frac{Lxm}{1.257xS}$$

### Formula 24.4

$$m = 1.257 \times \frac{Sxn^2}{L} \times 10^{-8}$$

## Application Example:

Calculate the inductance of a solenoid formed by 10 turns of wire with a length of 2 cm and a cross-sectional area of 2 cm².

NOTE: See resistance of a wire to calculate the cross-sectional area of a turn when the diameter is given.

Data:

$n = 10$

$S = 2 \text{ cm}^2$

$m = 2 \text{ cm}$

Applying Formula 24.1:

$$L = 1.257 \times \frac{2x10^2}{2} \times 10^{-8} = 1.257 \times 100 \times 10^{-8} = 125.7 \times 10^{-6}$$

$$L = 125.7 \ \mu H$$

# 25. INDUCTANCES IN SERIES

**(without magnetic coupling)**

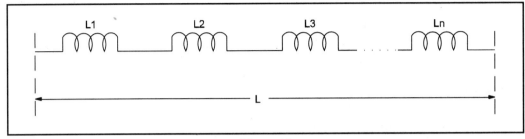

*Figure 18*

## Formula 25.1

$$L = L1 + L2 + L3 + \ldots\ldots\ldots + Ln$$

Where:

L is the equivalent inductance of the association in henrys (H)

L1, L2,. L3....Ln are the inductances of the associated coils in henrys (H)

**Particular Case:**

If the associated coils (n) have the same inductance the formula becomes:

## Formula 25.2

$$L = n \times L1$$

Where:

L is the equivalent inductance in henrys (H)

n is the number of associated coils in henrys (H)

L1 is the value of one of the associated coils in henrys (H)

**Application Example:**

Coils with inductances of 10 mH, 30 mH and 50 mH are wired in series. What is the equivalent inductance?

NOTE: Henry's submultiples can be used, but be sure that all the elements are expressed in the same submultiple and the result also.

Data:

L1 = 10 mH

L2 = 30 mH

L3 = 50 mH

Applying Formula 25.1:

L = 10 + 30 + 50 = 90 mH

# 26. INDUCTANCES IN PARALLEL

**(No magnetic coupling between coils)**

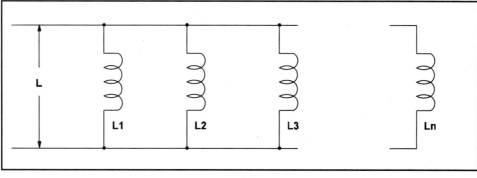

*Figure 19*

**Formula 26.1**

$$\frac{1}{L} = \frac{1}{L1} + \frac{1}{L2} + \frac{1}{L3} + \text{.......} \ \frac{1}{Ln}$$

Where:

L is the equivalent inductance in henrys (H)

L1, L2, L3......Ln are the associated inductances in henrys (H)

**Particular Case:**

If two coils are wired in parallel, calculations can be reduced by using the next formula:

**Formula 26.2**

$$\mathbf{L} = \frac{L1 x L2}{L1 + L2}$$

**Application Example:**

What is the equivalent inductance of two coils, one with an inductance of 40 mH and the other 60 mH, wired in parallel?

Data:

L1 = 40 mH

L2 = 60 mH

L = ?

Using Formula 26.2:

$$L = \frac{40 x 60}{40 + 60} = \frac{2400}{100} = 24 \text{ mH}$$

# 27. MUTUAL INDUCTANCE

If two coils are direct-coupled as shown in the figure below the mutual inductance depends on coupling coefficient (k), and the next formulas can be used.

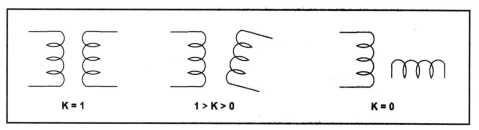

*Figure 21*

## Formula 27.1

$$M = k \times \sqrt{L1xL2}$$

Where:

M is the mutual inductance in henrys (H)

k is the coupling coefficient (between 0 and 1 as shown in the figure)

L1, L2 are the inductances of the coupled coils in henrys (H)

NOTE: The coupling coeffient is related to the amount of magnetic field lines produced by one coil that cuts the other coil in an angle that causes induction of an emf.

## Derivated Formulas:

## Formula 27.2

$$k = \frac{M}{\sqrt{L1xL2}}$$

# Part 2

# AC Formulas

# 28. FREQUENCY AND PERIOD

The frequency is numerically the number of cycles **per second** (Hz) of a periodic signal or wave and is in inverse proportion with the period (s).

**Formula 28.1**

$$f = \frac{1}{T}$$

Where:

f is the frequency in hertz (Hz)

T is the period in seconds (s)

**TABLE 14**

| If the period is in | the frequency is in |
|---|---|
| seconds (s) | hertz (Hz) |
| milliseconds (ms) | kilohertz (kHz) |
| microseconds (us) | megahertz (MHz) |
| nanoseconds (ns) | gigahertz (GHz) |

**Derivated Formula:**

**Formula 28.2**

$$T = \frac{1}{f}$$

**Application Example:**

Calculate the frequency of a signal with a period of 0.01 ms.

Data:

   $T = 0.01$ ms

   $f = ?$

Applying Formula 28.1:

$$f = \frac{1}{0.01} = 100 \text{ kHz} \quad \text{(period must be in miliseconds for result in kilohertz)}$$

# 29. CYCLIC OR ANGULAR FREQUENCY

The angular frequency is measured in radians per second (rd/s)

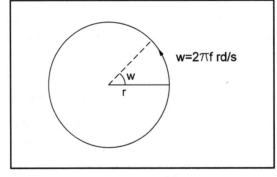

*Figure 22*

**Formula 29.1**

$$\omega = 2\pi f$$

Where:

   $\omega$ is the angular frequency in radians per second (rd/s)

   $\pi$ is the constant $= 3.1416$

   $f$ is the frequency in hertz (Hz)

**Derivated Formula:**

**Formula 29.2**

$$f = \frac{\omega}{2\pi}$$

**Application Example**:

Calculate the angular frequency of a 100 Hz alternating electric current.

Data:

f = 100 Hz

$\omega$ = ?

Using Formula 29.1:

$\omega$ = 2 x 3.14 x 100 = 628 rd/s

# 30. AVERAGE VALUE

To voltages in a sine wave this the simple average (arithmetic mean) of all the instantaneous values in one cycle, disregarding sign.

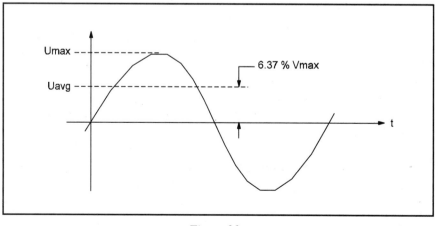

*Figure 23*

## Formula 30.1

$$U_{avg} = 0.637 \times U_{max}$$

Where:

$U_{max}$ is the highest positive or negative value of the voltage in one cycle in volts (V)

$U_{avg}$ is the average value in volts (V)

NOTE: The same concepts can by applied to currents and other periodic quantities with a sinewave shape.

## Derivated Formula:

## Formula 30.2

$$U_{max} = 1.567 \times U_{avg}$$

**Application Example:**

Calculate the peak or highest value of sine wave current with an average value of 100 volts.

Data:

Uavg = 100 V

Umax = ?

Using Formula 30.2:

Umax = 1.567 x 100 = 156.7 volts

# 31. RMS VALUE

The rms value, or root mean square value, is also called *effective* value and is equivalent to the same-numbered DC value in the heating effect it creates in a resistance.

For sine wave signals the rms value of a voltage can be calculated by the next formula:

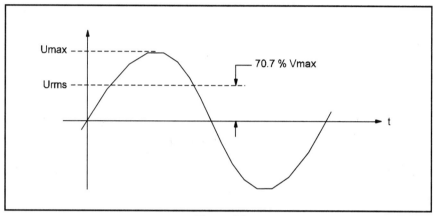

*Figure 24*

## Formula 31.1

$$U_{rms} = 0.707 \times U_{max}$$

Where:

Urms is the effective value of the voltage in volts (V)

Umax is the maximum value or peak value in volts (V)

NOTE: $0.707 = \sqrt{2}$

## Derivated Formula:

## Formula 31.2

$$U_{max} = 1.4142 \times U_{rms}$$

**Application Example:**

Determine the peak or maximum value of an AC sine 117 VAC voltage.

Data:

Urms = 117 V

Umax = ?

Applying Formula 31.2:

Umax = 1.4142 x 117 = 165.46 V

## TABLE 15
## Conversion Factors or Multiplier for Converting Maximum, Average and rms Values

Uavg = 0.637 Umax = 0.901 Urms

Urms = 0.707 Umax = 1.11 Uavg

Umax = 1.4142 Urms = 1.57 Uavg

- The same formulas can be applied to sinewave currents or other sinewave quantities.
- Peak-to-peak values (Upp) are defined as 2 x Umax.

# 32. FREQUENCY & WAVELENGTH

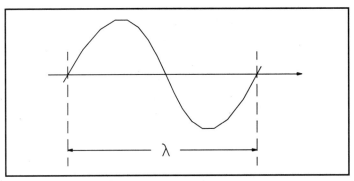

*Figure 25*

## Formula 32.1

$$v = \lambda \; x \; f$$

Where:

v is the propagation speed of the considered signal (in m/s)

$\lambda$ is the wavelength in meters (m)

f is the frequency in hertz (Hz)

NOTE: The formula can also be applied to light, radio waves and sound.

To sound: v = 340 m/s (average)

To light and radio waves: v = 300, 000, 000 m/s

## TABLE 16
## Velocity of Sound in Solids at 20°C

The velocities are considered to infinite medium. In rods the values are lower.

| Material | Velocity of longitudinal waves (m/s) | Velocity of transverse waves (m/s) |
|---|---|---|
| Aluminium | 5 080 | 3 080 |
| Brass | 3 490 | 2 123 |
| Copper | 3 710 | 2 270 |
| Glass | 3 400 to 5 300 | 2 000 to 3 500 |
| Ice | 3 280 | 1 990 |
| Iron | 5 170 | 3 230 |
| Lead | 2 640 | 1 590 |
| Marble | - | 3 260 |
| Mica | - | 2 160 |
| (continued next page) | | |

| | | |
|---|---|---|
| Nickel | 4 785 | 2 960 |
| Porcelain | 1 884 | 3 120 |
| Rubber | 46 | 27 |
| Steel | 5 053 | 3 300 |
| Tin | 2 730 | 1 670 |
| Tungsten | 4 310 | 2 620 |
| Zinc | 3 810 | 2 410 |

NOTE: The values depend on the degree of purity of the material and also, in the case of alloys, the particular composition of the material.

## TABLE 17

### Velocity of Sound in Gas (1 atm x 0°C)

| Gas | Velocity (m/s) |
|---|---|
| Air | 331 |
| Ammonia | 415 |
| Carbon dioxide | 259 |
| Deuterium | 890 |
| Helium | 965 |
| Hydrogen | 1 284 |
| Neon | 435 |
| Nitrogen | 334 |
| Oxygen | 316 |

### Formula 32.2

$$f = \frac{v}{\lambda}$$

## Formula 32.3

$$\lambda = \frac{v}{f}$$

**Application Example:**

What is the wavelength of a 150 MHz radio signal?

Data:

f = 150 MHz = 150, 000, 000 Hz

v = 300, 000, 000 m/s (radiowaves)

$\lambda$ = ?

Applying Formula 32.3

$\lambda$ = 300, 000, 000/150, 000, 000 = 2 meters

# 33. CAPACITIVE REACTANCE

The opposition offered by a capacitance to an AC current is termed capacitive reactance and depends on the capacitance and also the AC frequency. The next formula is used to calculate the capacitive reactance of a capacitor in an AC circuit.

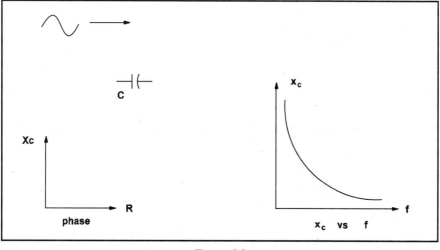

*Figure 26*

## Formula 33.1

$$Xc = \frac{1}{2 \times \pi \times f \times C}$$

Where:

Xc is the capacitive reactance in ohms ($\Omega$)

f is the frequency in hertz (Hz)

$\pi$ is the constant 3.1416

C is the capacitance in farads (F)

NOTE: A pure capacitance consumes no power, as the power stored in the magnetic field of the capacitor in one quarter-cycle is returned to the circuit at the following quarter-cycle.

## Derivated Formulas:

## Formula 33.2

$$C = \frac{1}{2 \times \pi \times f \times Xc}$$

## Formula 33.3

$$f = \frac{1}{2 \times \pi \times C \times Xc}$$

**Application Example:**

A 0.2 $\mu$F capacitor is operating in a 100 kHz circuit. Calculate reactance at that frequency.

Data:

C = 0.2 uF = 0.2 x $10^{-6}$ F

f = 100 kHz = $10^5$ Hz

Xc = ?

Applying Formula 33.1:

$$Xc = \frac{1}{2 x 3.14 x 10^{-6} x 10^{5}} = \frac{1}{0.628} = 1.59 \text{ ohms}$$

# 34. INDUCTIVE REACTANCE

The oposition offered by an inductance to an AC current depends on the reactance value and also the AC frequency. The next formula is used to calculate the inductive reactance (Xc).

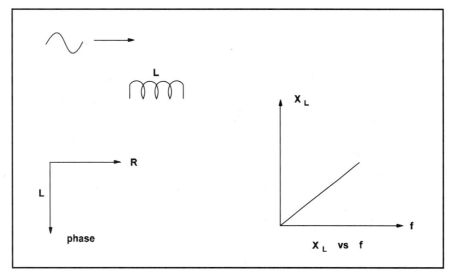

*Figure 27*

## Formula 34.1

$$\mathbf{X_L} = 2 \text{ x } \pi \text{ x f x C}$$

Where:

XL is the inductive reactance in ohms ($\Omega$)

$\pi$ is the constant 3.1416

f is the frequency in hertz (Hz)

C is the capacitance in farads (F)

## Derivated Formulas:

### Formula 34.2

$$f = \frac{XL}{2 x \pi x L}$$

### Formula 34.3

$$L = \frac{XL}{2 x \pi x f}$$

**Application Example:**

Through a 2 mH inductor flows a 100 mA, 100 Hz sinewave current. Assuming that it is a pure inductance, calculate the inductive reatance.

Data:

$$L = 2\,mH = 2 \times 10^{-3}\,H$$

$$f = 100\,Hz$$

$$XL = ?$$

Using Formula 34.1:

$$XL = 2 \times 3.14 \times 100 \times 10^{-3} = 0.628\,ohms$$

NOTE: Some multiples and submultiples of the units can be used as shown in the following table:

**TABLE 18**

Using:

| Inductance in | Frequency in | You'll find the reactance in: |
|---|---|---|
| Henry | Hertz | Ohm |
| Milihenry | Hertz | Miliohm |
| Microhenry | Hertz | Microhm |
| Henry | Kilohertz | Kilohm |
| Henry | Megahertz | Megohm |
| Milihenry | Kilohertz | Ohm |
| Microhenry | Kilohertz | Miliohm |
| Microhenry | Megahertz | Ohm |

# 35. QUALITY FACTOR (FACTOR-Q)

The figure of merit or quality factor of an AC device is termed Q. The Q is associated to the selectivity of a circuit. The higher the Q better the circuit capacities in separate near-frequency signals. This factor is defined by next formula:

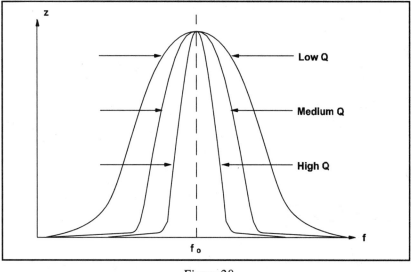

*Figure 28*

## Formula 35.1

$$Q = \frac{XL}{R}$$

Where:

Q is the quality factor

XL is the inductive reactance of the coil in ohms ($\Omega$)

R is the losses in the coil resistance, or the ohmic resistance in ohms ($\Omega$)

NOTE: The same concept can be applied to any AC device when the factor XL becomes X (the total reactance in ohms).

## Derivated Formulas:

## Formula 35.2

$$Q = \frac{\omega x L}{R}$$

Where:

$\omega$ is 2 x 3.14 x f (f is the frequency in hertz)

## Formula 3 5.3

$$R = \frac{X_L}{Q}$$

## Formula 35.4

$$X_L = Q \times R$$

**Application Example:**

A 2 mH choke has a resistance of 10 ohms and a 1 MHz impedance of 500 ohms. Calculate the Q of this choke at the given frequency.

Data:

$X_L$ = 500 ohms

R = 10 ohms

Q = ?

Using Formula 35.1:

Q = 500/10 = 50

# 36. OHM'S LAW FOR AC CIRCUITS

Altough Ohm's Law in the original form is applied to DC circuits, it applies to AC as well, as long as the resistance is considered pure. Ohm's Law for AC circuits is often written with Z replacing the R as given in the next formula:

**Formula 36.1**

## Urms = Z x Irms

Where:

Urms = rms voltage across the circuit in volts (V)

Z = circuit impedance in ohms ($\Omega$)

Irms = rms current through the circuit in amperes (A)

**Derivated Formulas:**

**Formula 36.2**

$$Irms = \frac{Urms}{Z}$$

## Formula 36.3

$$Z = \frac{Urms}{Irms}$$

**Application Example:**

Calculate the current flowing in a circuit with an impedance of 100 ohms when connected to a 117Vrms circuit.

Data:

>    Urms = 117 V

>    Z = 100 ohms

Using Formula 36.2:

>    Irms = 117/100 = 1.17 A

# 37. RL IN SERIES

To calculate the impedance of an RL series circuit the next formula is used:

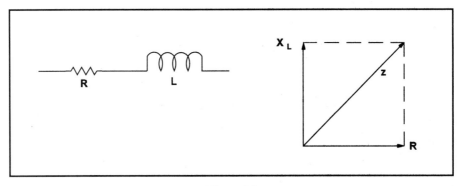

*Figure 29*

## Formula 37.1

$$Z = \sqrt{X_L^2 + R^2}$$

Where:

Z is the circuit impedance in ohms ($\Omega$)

$X_L$ is the inductve reactance in ohms ($\Omega$)

R is the resistance in ohms ($\Omega$)

## Derivated Formulas:

## Formula 37.2

$$X_L = \sqrt{Z^2 - R^2}$$

## Formula 37.3

$$R = \sqrt{Z^2 - X_L^2}$$

## Application Example:

What is the impedance of a circuit formed by a 20-ohm resistor in series with a coil with an impedance of 30 ohms in the frequency of operation?

Data:

R = 20 ohms

$X_L$ = 30 ohms

Applying Formula 37.1:

$$Z = \sqrt{20^2 + 30^2} = \sqrt{400 + 900} = \sqrt{1300} = 36.05 \text{ ohms}$$

# 38. RC IN SERIES

The impedance of an RC series circuit is calculated by the next formula:

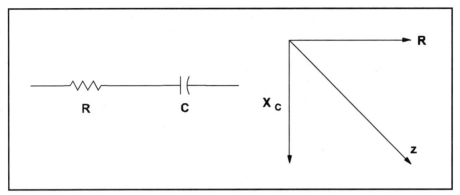

*Figure 30*

## Formula 38.1

$$Z = \sqrt{Xc^2 + R^2}$$

Where:

Z is the impedance in ohms ($\Omega$)

Xc is the capacitive reactance in ohms ($\Omega$)

R is the resistance in ohms ($\Omega$)

## Derivated Formulas:

## Formula 38.2

$$Xc = \sqrt{Z^2 - R^2}$$

## Formula 38.3

$$R = \sqrt{Z^2 - Xc^2}$$

**Application Example:**

Determinate the impedance of a 0.1 $\mu$F capacitor in series with a 10-ohm resistor in a circuit where the capacitive reactance of the capacitor is 30 ohms.

Data:

Xc = 30 ohms

R = 10 ohms

Applying Formula 38.1:

$$Z = \sqrt{30^2 + 10^2} = \sqrt{900 + 100} = \sqrt{1000} = 31.62 \text{ ohms}$$

# 39. LC IN SERIES

The following formula is used to find the impedance of a capacitor and an inductor wired in series.

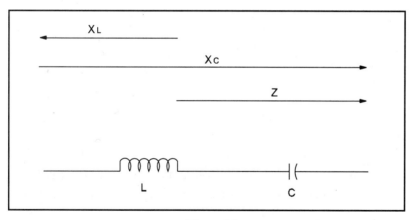

*Figure 31*

**Formula 39.1**

$$Z = Xc - X_L$$

Where:

> Z is the equivalent impedance in ohms ($\Omega$)

> Xc is the capacitive reactance in ohms ($\Omega$)

> $X_L$ is the inductive reactance in ohms ($\Omega$)

## NOTE:

- If Xc is larger than $X_L$ the circuit has a capacitive characteristic

- If Xc is smaller than $X_L$ consider the absolute value (without the minus sign) and the circuit has a inductive characteristic.

- If Z=0 the circuit is resonant in the operation frequency (see resonance formula for more details).

## Derivated Formulas:

### Formula 39.2

$$Xc = X_L + Z$$

### Formula 39.3

$$X_L = Xc + Z$$

**Application Example:**

A capacitor with a capacitive reactance of 3 ohms is wired in series with a coil having an inductive reactance of 4 ohms in the same circuit. Calculate the impedance presented by this association.

Data:

> Xc = 3 ohms

> $X_L$ = 4 ohms

> Z = 3 - 4 = -1 ohms (inductive impedance)

# 40. RLC IN SERIES

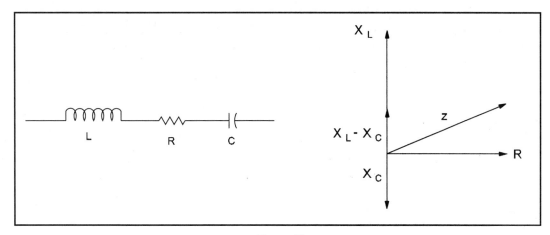

*Figure 32*

## Formula 40.1

$$Z = \sqrt{R^2 + (X_L - Xc)^2}$$

Where:

Z is the equivalent impedance in ohms ($\Omega$)

R is the resistance in ohms ($\Omega$)

$X_L$ is the associated inductive reactance in ohms ($\Omega$)

Xc is the associated capacitive reactance in ohms ($\Omega$)

## Derivated Formulas:

## Formula 40.2

$$R = \sqrt{Z^2 - (X_L - Xc)^2}$$

## Formula 40.3

$$X_L - Xc = \sqrt{Z^2 - R^2}$$

**Application Example:**

Calculate the impedance of a series circuit formed by a 10-ohm resistor and a capacitor with a capacitive reactance of 5 ohms and a coil with an inductive reactance of 10 ohms.

Data:

$R = 10$ ohms

$Xc = 5$ ohms

$X_L = 10$ ohms

Applying Formula 40.1:

$$Z = \sqrt{10^2 + (10-5)^2} = \sqrt{100 + 25} = \sqrt{125} = 11.18 \text{ ohms}$$

# 41. RC IN PARALLEL

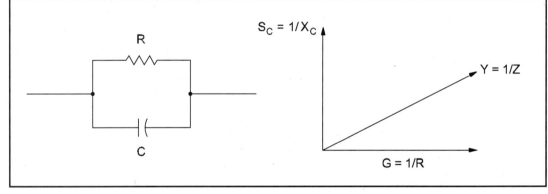

*Figure 33*

## Formula 41.1

$$Z = \frac{R x Xc}{\sqrt{R^2 + Xc^2}}$$

Where:

Z is the impedance in ohms ($\Omega$)

R is the resistance in ohms ($\Omega$)

Xc is the capacitive reactance in ohms ($\Omega$) .

NOTE: In the vectorial representation we defined two new important quantities:

- Y (the inverse of impedance) is named "admittance" and is measured in siemens (S)
- G (the inverse of resistance) is named "conductance" and is expressed in siemens (S)
- $S_L$ and Sc (the inverse of reactances) are named "susceptance," measured in siemens (S)

**Application Example:**

Calculate the impedance of a capacitor and a resistor wired in parallel. As given, the capacitive reactance of the capacitor is 20 ohms and the resistance is 10 ohms.

Data:

R = 10 ohms

Xc = 20 ohms

Applying Formula 41.1:

$$Z = \frac{10 x 20}{\sqrt{10^2 + 20^2}} = \frac{200}{\sqrt{500}} = \frac{200}{22.36} = 8.94 \text{ ohms}$$

# 42. LR IN PARALLEL

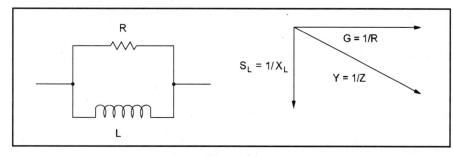

*Figure 34*

**Formula 42.1**

$$Z = \frac{R x X_L}{\sqrt{R^2 + X_L^2}}$$

Where:

Z is the impedance in ohms ($\Omega$)

R is the resistance in ohms ($\Omega$)

$X_L$ is the inductive reactance in ohms ($\Omega$)

# 43. LC IN PARALLEL

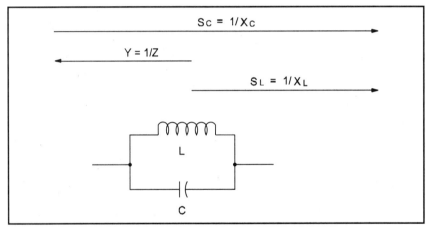

*Figure 35.*

**Formula 43.1**

$$Z = \frac{X_L x XC}{X_L - Xc}$$

Where:

Z is the impedance in ohms ($\Omega$)

$X_L$ is the inductive reactance in ohms ($\Omega$)

Xc is the capacitive reactance in ohms ($\Omega$)

NOTE: If the result is positive, the circuit is inductive; if the result is negative the circuit is capacitive.

**Application Example:**

Determine the impedance of a circuit formed by a capacitor and an inductor wired in paralell. The capacitor has a capacitive reactance of 5 ohms and the inductor 8 ohms in the indicated circuit.

Data:

$$Xc = 5 \text{ ohms}$$

$$X_L = 8 \text{ ohms}$$

$$Z = ?$$

Applying Formula 43.1:

$$Z = \frac{8x5}{8-5} = \frac{40}{3} = 13.33 \text{ ohms (inductive)}$$

# 44. RLC IN PARALLEL

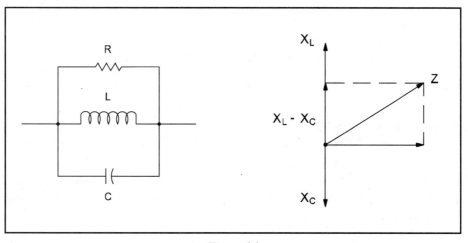

*Figure 36*

**Formula 44**

$$Z = \cfrac{1}{\sqrt{\left(\dfrac{1}{R}\right)^2 + \left(\dfrac{X_L - Xc}{X_L xXc}\right)^2}}$$

Where:

Z is the impedance in ohms ( $\Omega$ )

R is the resistance in ohms ( $\Omega$ )

$X_L$ is the inductive reactance in ohms ( $\Omega$ )

XC is the capacitive reactance in ohms ( $\Omega$ )

**Derivated Formula:**

**Formula 44.2**

$$Y = \sqrt{G^2 + \left(S_L - S_C\right)^2}$$

Where:

Y is the admittance (inverse of impedance) in siemens (S)

G is the conductance (inverse of resistance) in siemens (S)

$S_L$ is the inductive susceptance (inverse of reactance) in siemens (S)

Sc is the capacitive susceptance in siemens (S)

# 45. RESONANCE (LC)

The frequency in which the capacitive reactance of an LC circuit becomes equal to the inductive reactance is termed "resonant frequency," and both for parallel and series circuits can be calculated by the following formula:

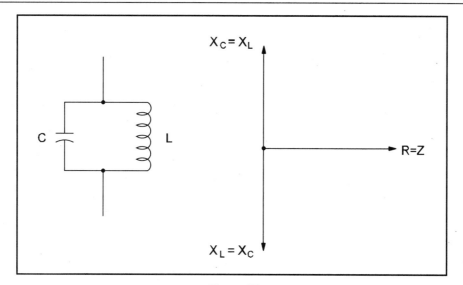

*Figure 37*

## Formula 45.1

$$fr = \frac{1}{2 \times \pi \times \sqrt{L \times C}}$$

Where:

f is the resonant frequency in hertz (Hz)

$\pi = 3.1416$

L is the inductance in henry (H)

C is the capacitance in farads (F)

NOTE: You can use multiples and submultiples of the units for capacitance and inductance finding results in multiples of hertz according to the next table:

| Using inductance in | And capacitance in | You'll find the frequency in |
|---|---|---|
| Henry | Farad | Hertz |
| Miicrohenry | microFarad | Megahertz |

## Derivated Formula:

### Formula 45.2

$$L = \frac{1}{\omega^2 xC}$$

Where:

$\omega = 2 \times \Pi \times f$  (f is the frequency in Hz)

### Formula 45.3

$$C = \frac{1}{\omega^2 xL}$$

Where:

$\omega = 2 \times \pi \times f$  (f is the frequency in Hz)

### Formula 45.4

When capacitance is in pF, inductance in $\mu$ H and frequency in MHz, the following formula can be used:

$$f = \frac{159}{\sqrt{LxC}}$$

### Formula 45.5

When the capacitance is in pF, inductance in mH the frequency is in kHz and the following formula can be used:

$$f = \frac{5,033}{\sqrt{LxC}}$$

Important Properties:

- In the ideal series circuit at the resonant frequency the impedance is infinite.
- In the ideal parallel circuit at the resonant frequency the impedance is zero.

**Application Example:**

Determine the resonant frequency of a 10 uH coil wired in parallel with a 100 pF capacitor.

Data:

L = 10 uH

C = 100 pF

Using Formula 45.4:

$$f = \frac{159}{\sqrt{10x100}} = \frac{159}{\sqrt{1000}} = \frac{159}{31.62} = 5.028 \text{ MHz}$$

# 46. TIME CONSTANT (RC circuit)

The time constant of an RC circuit is the time interval needed to either charge the capacitor via a resistor to 63.2 % of the total charge, or discharge from the total charge to 37.8 %.

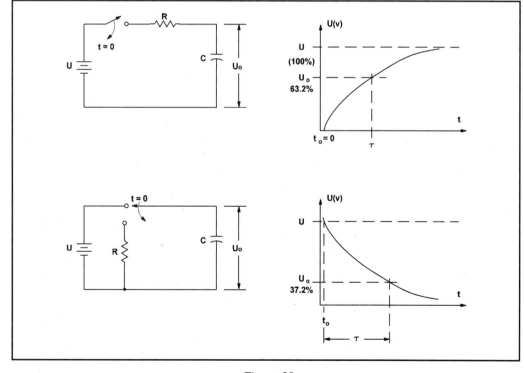

*Figure 38*

## Formula 46.1

$$\tau = R \times C$$

Where:

$\tau$ is the time constant in seconds (s)

R is the resistance in ohms ($\Omega$)

C is the capacitance in Farads (F)

## TABLE 19

NOTE: If multiples and submultiples are used the next table is useful:

| When using R in | And C in | $\tau$ will be found in |
|---|---|---|
| Ohms | Farads | Seconds |
| Ohms | Microfarads | Microseconds |
| Kilohms | Farads | Kiloseconds |
| Kilohms | Microfarads | Milliseconds |
| Megohms | Microfarads | Seconds |

## Derivated Formulas:

## Formula 46.2

$$R = \frac{\tau}{C}$$

## Formula 46.3

$$C = \frac{\tau}{R}$$

**Application Example:**

Determine the time constant of a circuit formed by a 100 kohm resistor in series with a 500 microfarad capacitor. Calculate the voltage across the capacitor after the considered time interval when the circuit is subjected to a 100V power supply.

Data:

$R = 100 \times 10^3$ ohms

$C = 500 \times 10^{-6}$ Farads

$\tau = ?$

Using Formula 46.1:

$\tau = 100 \times 10^3 \times 500 \times 10^{-6} = 50$ seconds

Final Voltage:

$U = 0.63 \times 100 = 63.2V$

# 47. TIME CONSTANT (LC)

The interval between the instant in which the current begins to flow for an inductor in series with a resistor and the instant in which the current is 63.2% of the final value is the time constant of the RL series circuit. It is also the value between the instant in which the circuit is open and the instant in which the current decays to 37.8% of maximum.

## Formula 47.1

$$\tau = \frac{L}{R}$$

Where:

$\tau$ is the time constant in seconds (s)

L is the inductance in henrys (H)

R is the resistance in ohms ($\Omega$)

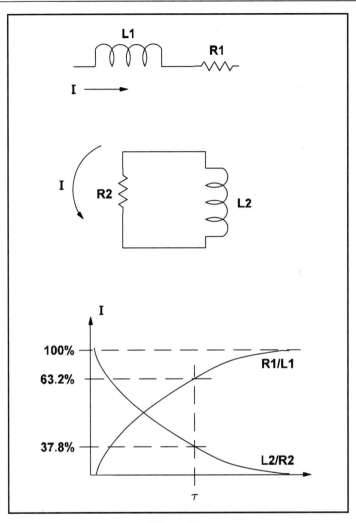

*Figure 39*

## Derivated Formulas:

**Formula 47.2**

$$L = \tau \times R$$

**Formula 47.3**

$$R = \frac{L}{\tau}$$

**Application Example:**

What is the time constant of a circuit formed by a 20 kohm resistor in series with a 10 mH inductor?

Data:

$R = 20 \text{ k}\Omega = 20\,000\ \Omega$

$L = 10 \text{ mH} = 0.01 \text{ H}$

$\tau = ?$

Using Formula 47.2:

$\tau = 20,000 \times 0.01 = 200$ seconds

# 48. INDUCTIVE COUPLING USING TRANSFORMERS

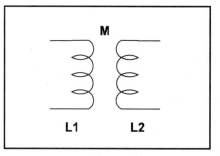

*Figure 40*

**Formula 48.1**

$$K = \frac{M}{\sqrt{L1xL2}}$$

Where:

K is the coupling factor

M is the mutual inductance in henrys (H)

# 49. DIRECT INDUCTIVE COUPLING

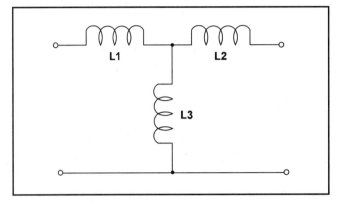

*Figure 41*

## Formula 49.1

$$K = \frac{L3}{\sqrt{(L1+L3)x(L2+L3)}}$$

Where:

L1, L2, L3 are the used inductances in henrys (H)

K is the coupling factor

NOTE: The inductances are not magnetic-coupled.

# 50. OHMIC COUPLING

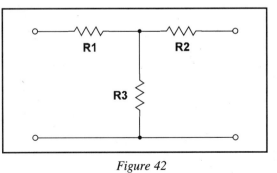

*Figure 42*

**Formula 50.1**

$$K = \frac{R3}{\sqrt{(R1+R3)x(R2+R3)}}$$

# 51. CAPACITIVE COUPLING

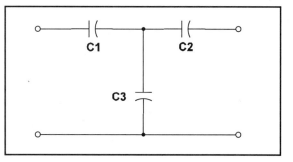

*Figure 43*

**Formula 51.1**

$$K = \frac{\sqrt{C1xC2}}{\sqrt{(C1+C3)x(C2+C3)}}$$

Where:

K is the coupling factor

C1, C2, C3 are the involved capacitances in farads (F)

NOTE: Any submultiple of farad can be used if all the capacitances are equally expressed.

# 52. LOW-PASS FILTERS

A Low-pass filter is an LC network designed to let frequencies under a certain value pass without attenuation, and blocking frequencies above this value.

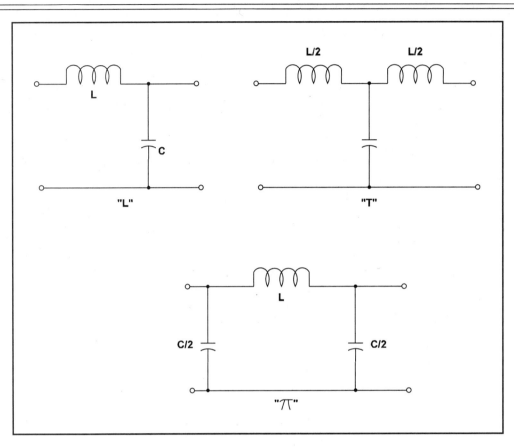

*Figure 44*

## Formula 52.1

$$fc = \frac{1}{\pi x \sqrt{CxL}}$$

Where:

fc is the cutoff frequency in hertz (Hz)

π is the constant 3.1416

C is the capacitance in farads (F)

L is the inductance in henrys (H)

**Application Example:**

Calculate the cutoff frequency of an LC low-pass filter formed by a 1 mH inductor in series with a 1000 $\mu$ F capacitor.

Data:

$C = 1\,000\ \mu F = 10^{-3}\ F$

$L = 1\ mH = 10^{-3}\ H$

$fc = ?$

Appying Formula 52.1:

$$fc = \frac{1}{2x3.14x\sqrt{10^{-3}x10^{-3}}} = \frac{1}{6.28x\sqrt{10^{-6}}} = \frac{10^3}{6.28} = 159.23\ Hz$$

# 53. HIGH-PASS FILTERS

A high-pass filter is a network formed by capacitors and coils designed to let frequencies above a certain value pass without attenuation, but blocking frequencies under that value. Depending on the number of coils and capacitors, the high-pass filter can have different designations as shown in the figure.

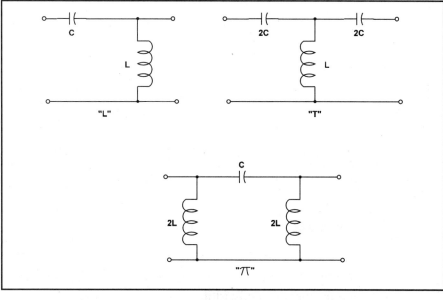

*Figure 45*

## Formula 53.1

$$fc = \frac{1}{4 \times \pi \times \sqrt{L \times C}}$$

Where:

fc is the cutoff frequency in hertz (Hz)

$\pi$ is the constant 3.1416

L is the inductance used in the circuit shown above in henry (H)

C is the capacitance used in the circuit shown above in farad (F)

NOTE: The calculated value Fc is the lower cutoff frequency, as the upper cutoff frequency is $\infty$.

**Application Example:**

Calculate the cutoff frequency of a high-pass filter formed by a 100 nF capacitor and a 1 mH inductor (L-filter).

Data:

$C = 100 \text{ nF} = 100 \times 10^{-9} \text{ F}$

$L = 1 \text{ mH} = 10^{-3} \text{ H}$

$fc = ?$

Applying Formula 53.1:

$$fc = \frac{1}{4 \times 3.14 \times \sqrt{10^{-3} \times 10^{-9}}} = \frac{1}{12.56 \times \sqrt{10^{-12}}} = \frac{10^6}{12.56} = 0.0796 \text{ MHz or } 79.6 \text{ kHz}$$

# 54. BAND-PASS FILTERS

These are networks formed by coils and capacitors as shown below. These circuits present a low impedance to a certain frequency bandwidth.

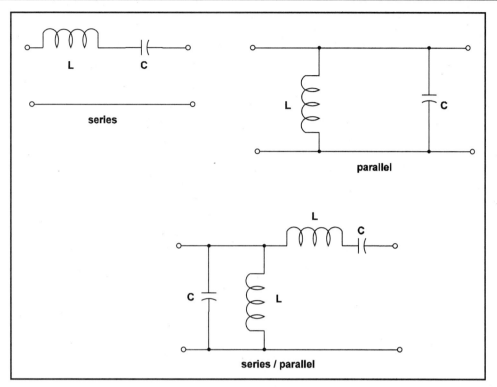

*Figure 46*

## Formula 54.1

$$fc = \frac{1}{2 \times \pi \times \sqrt{L \times C}}$$

Where:

fc is the tune frequency or central frequency in hertz (Hz)

π is the constant 3.1416

L is the inductance in henry (H)

C is the capacitance in farad (F)

# 55. DIFFERENTIATION

The circuit shown below is used for differentiation of a signal. The output is function of the input signal, time and components used, calculated by the following formula:

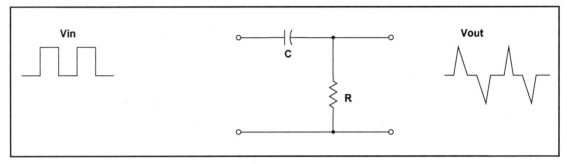

*Figure 47*

## Formula 55.1

$$\text{Uout} = \text{R x C x } \frac{dUin}{dt}$$

Where:

Uout is the output voltage in volts (V)

Uin is the input voltage in volts (V)

R is the resistance in ohms ($\Omega$)

C is the capacitance in farads (F)

## Derivated Formulas:

Considering that $\tau = \text{R x C}$ we can also write:

$$\text{Uout} = \tau \text{ x R x C}$$

# 56. INTEGRATION

The output voltage as function of time is given by the next formula in the integration circuit.

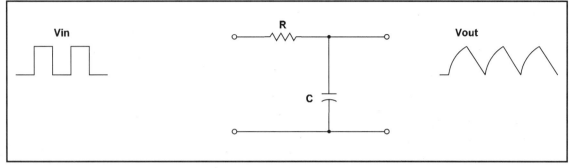

*Figure 48*

## Formula 56.1

$$\mathbf{Uout} = \frac{1}{RxC}\int Uinxdt$$

Where:

Uout is the output voltage in volts (V)

Uin is the input voltage in volts (V)

R is the resistance in ohms ($\Omega$)

C is the capacitance in farads (F)

# 57. NOISE

Thermal noise is generated by the conversion of electric energy into electromagnetic energy (radio, IF, light and UV radiation). This can be associated to resistance and is calculated by the next formula:

## Formula 57.1

$$Pn = 4 \times k \times T \times \Delta f$$

Where:

Pn is the power noise in watts (W)

k is Boltzmann's Constant = 1.38 x 10$^{-23}$ Ws/$^{\circ}$K

T is the temperature in degrees kelvin ($^{\circ}$K)

$\Delta f$ is the bandwidth in hertz (Hz)

# 58. BANDWIDTH

The bandwidth of a circuit is given by the difference between the maximum and minimum frequencies, where the impedance isn't less than 70.7% of the maximum value.

The bandwidth, as shown in the figure below, depends on the frequency and also the merit figure (Q).

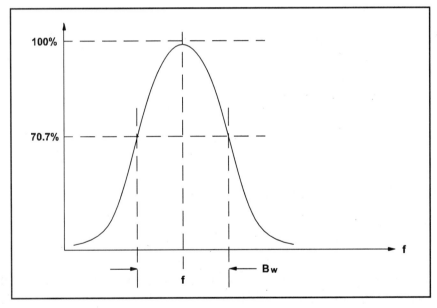

*Figure 49*

## Formula 58.1

$$Bw = \frac{f}{Q}$$

Where:

Bw is the bandwidth in hertz (Hz)

f is the central or resonant frequency in hertz (Hz)

Q is the merit figure or Q-factor

**Application Example:**

Calculate the bandwidth of a resonant circuit operating at a frequency of 1 MHz with a merit figure of 200.

Data:

f = 1 MHz = 1 000 000 Hz

Q = 200

Bw = ?

Using Formula 58.1:

$$Bw = \frac{1000000}{200} = 5\ 000\ Hz = 5\ kHz$$

# 59. VOLTAGE RATIO IN TRANSFORMERS

The voltage ratio in transformer's windings depends on the number of turns of each winding and is calculated by the following formula:

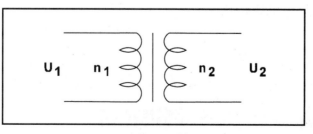

*Figure 50*

## Formula 59.1

$$\frac{U1}{U2} = \frac{n1}{n2}$$

Where:

U1 is the voltage applied to the primary winding in volts (V)

U2 is the voltage in the secondary winding in volts (V)

n1 is the number of turns in the primary winding

n2 is the number of turns in the secondary winding

**Application Example:**

A transformer has a primary winding formed by 100 turns and a secondary winding formed by 50 turns. Calculate the secondary voltage when applying 120V to the primary.

Data:

U1 = 120V (AC)

n1 - 100 turns

n2 = 50 turns

U2 = ?

Applying Formula 59.1:

$$\frac{120}{U2} = \frac{100}{50}$$

$$100 \times U2 = 120 \times 50$$

$$U2 = \frac{120x50}{100} = 60 \text{ volts}$$

# 60. CURRENT RATIO IN TRANSFORMERS

The current ratio in a transformer's windings depends on the number of turns used in the windings according the following formula.

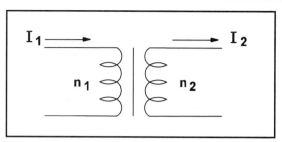

*Figure 51*

## Formula 60.1

$$\frac{I1}{I2} = \frac{n2}{n1}$$

Where:

I1 is the current flowing through the primary winding in amperes (A)

I2 is the current flowing through the secondary winding in amperes (A)

n1 is the number of turns of primary winding

n2 is the number of turns of secondary winding

**Application Example:**

Consider a transformer with a 100-turn primary and a 200-turn secondary. When a 100 mA load is powered from the secondary, calculate the current flowing through the primary.

Data:

n1 = 100 turns

n2 = 200 turns

I2 = 100 mA

I1 = ?

Applying Formula 60.1:

$$\frac{100}{I1} = \frac{200}{100}$$

I1 x 200 = 100 x 100

$$I1 = \frac{100x100}{200} = 50 \text{ mA}$$

## TABLE 20
## Impedance Ratio and Turns Ratio of Transformers

The impedance ratio is calculated by the formula:

$$Z1/Z2 = n1^2/n2^2$$

The next table gives the impedance ratio for some common values of turns ratios:

| Turns Ratio Ratio | Impedance Ratio | Turns Ratio | Impedance Ratio | Turns Ratio | Impedance Ratio |
|---|---|---|---|---|---|
| 100:1 | 10 000:1 | 85:1 | 7 225:1 | 70:1 | 4 900:1 |
| 99:1 | 9 801:1 | 84:1 | 7 056:1 | 69:1 | 4 761:1 |
| 98:1 | 9 604:1 | 83:1 | 6 889:1 | 68:1 | 4 624:1 |
| 97:1 | 9 409:1 | 82:1 | 6 724:1 | 67:1 | 4 489:1 |
| 96:1 | 9 216:1 | 81:1 | 6 571:1 | 66:1 | 4 356:1 |
| 95:1 | 9 025:1 | 80:1 | 6 400:1 | 65:1 | 4 225:1 |
| 94:1 | 8 836:1 | 79:1 | 6 241:1 | 64:1 | 4 096:1 |
| 93:1 | 8 649:1 | 78:1 | 6 084:1 | 63:1 | 3 969:1 |
| 92:1 | 8 464:1 | 77:1 | 5 929:1 | 62:1 | 3 844:1 |
| 91:1 | 8 281:1 | 76:1 | 5 776:1 | 61:1 | 3 721:1 |
| 90:1 | 8 100:1 | 75:1 | 5 625:1 | 60:1 | 3 600:1 |
| 89:1 | 7 921:1 | 74:1 | 5 476:1 | 59:1 | 3 481:1 |
| 88:1 | 7 744:1 | 73:1 | 5 329:1 | 58:1 | 3 364:1 |
| 87:1 | 7 569:1 | 72:1 | 5 184:1 | 57:1 | 3 249:1 |
| 86:1 | 7 396:1 | 71:1 | 5 041:1 | 56:1 | 3 136:1 |

(continued next page)

| Turns Ratio | Impedance Ratio | Turns Ratio | Impedance Ratio | Turns Ratio | Impedance Ratio |
|---|---|---|---|---|---|
| 55:1 | 3 025:1 | 36:1 | 1 296:1 | 17:1 | 289:1 |
| 54:1 | 2 916:1 | 35:1 | 1 225:1 | 16:1 | 256:1 |
| 53:1 | 2 809:1 | 34:1 | 1 156:1 | 15:1 | 225:1 |
| 52:1 | 2 704:1 | 33:1 | 1 089:1 | 14:1 | 196:1 |
| 51:1 | 2 601:1 | 32:1 | 1 024:1 | 13:1 | 169:1 |
| 50:1 | 2 500:1 | 31:1 | 961:1 | 12:1 | 144:1 |
| 49:1 | 2 401:1 | 30:1 | 900:1 | 11:1 | 121:1 |
| 48:1 | 2 304:1 | 29:1 | 841:1 | 10:1 | 100:1 |
| 47:1 | 2 209:1 | 28:1 | 784:1 | 9:1 | 81:1 |
| 46:1 | 2 116:1 | 27:1 | 729:1 | 8:1 | 64:1 |
| 45:1 | 2 025:1 | 26:1 | 676:1 | 7:1 | 49:1 |
| 44:1 | 1 936:1 | 25:1 | 625:1 | 6:1 | 36:1 |
| 43:1 | 1 849:1 | 24:1 | 576:1 | 5:1 | 25:1 |
| 42:1 | 1 764:1 | 23:1 | 529:1 | 4:1 | 16:1 |
| 41:1 | 1 681:1 | 22:1 | 484:1 | 3:1 | 9:1 |
| 40:1 | 1 600:1 | 21:1 | 441:1 | 2:1 | 4:1 |
| 39:1 | 1 521:1 | 20:1 | 400:1 | 1:1 | 1:1 |
| 38:1 | 1 444:1 | 19:1 | 361:1 | | |
| 37:1 | 1 369:1 | 18:1 | 324:1 | | |

# 61. DECIBEL

The Bel (B) is an unit used to express the ratio between two quantities of the same nature (power, current or voltage) in a logarithmic form.

The next formulas are used to calculate power, voltage and current ratios in decibels (1 dB = 0.1 B)

The decibel is derivated from common logarithms (base 10).

NOTE: Since the decibel is a comparasion unit (not an absolute value), in some cases a reference level must be indicated. When working with amplifiers, for instance, it is common to adopt as a reference or zero level 1 mW (600 ohms).

If 1 mW is adopted as reference, the letter "m" is added to dB, which becomes "dBm".

## Formula 61.1

$$dB = 10 \lg \frac{Pout}{Pin}$$

Where:

  dB is the power gain or loss (pure number)

  lg is the common logarithm of the expression (base 10)

  Pout is the output power in watts (W)

  Pin is the input power in watts (W)

## Formula 61.2

$$dB = 20 \lg \frac{Uout}{Uin}$$

Where:

  dB is the voltage gain or loss (pure number)

  lg is the common logarithm of the expression (base 10)

  Uout is the output voltage in volts (V)

  Uin is the input voltage in volts (V)

## Formula 61.3

$$dB = 20 \lg \frac{Iout}{Iin}$$

Where:

  dB is the current ratio (pure number)

  lg is the common logarithm of the expression (base 10)

  Iout is the output current in amperes (A)

  Iin is the input current in amperes (A)

**Application Example:**

An amplifier is said to have a power output of 10W when a 0.5W signal is applied to its input. Calculate the power gain in dB.

Data:

Pout = 10W

Pin = 0.5

dB = ?

Applying Formula 61.1:

$$dB = 20 \lg \frac{10}{0.5} = 20 \lg 2 = 20 \times 0.65 = 13 \text{ dB}$$

## TABLE 21
## Voltage or Current Ratios vs Ratios and Decibels

| Power Ratio | Voltage or Current Ratio | dB | Power Ratio | Voltage or Current Ratio | dB |
|---|---|---|---|---|---|
| 1.000 | 1.000 | 0 | 1.413 | 1.189 | 1.5 |
| 1.023 | 1.012 | 0.1 | 1.585 | 1.259 | 2.0 |
| 1.047 | 1.023 | 0.2 | 1.778 | 1.334 | 2.5 |
| 1.072 | 1.035 | 0.3 | 1.995 | 1.413 | 3.0 |
| 1.096 | 1.047 | 0.4 | 2.239 | 1.496 | 3.5 |
| 1.122 | 1.059 | 0.5 | 2.512 | 1.585 | 4.0 |
| 1.148 | 1.072 | 0.6 | 2.818 | 1.679 | 4.5 |
| 1.175 | 1.084 | 0.7 | 3.162 | 1.778 | 5.0 |
| 1.202 | 1.096 | 0.8 | 3.548 | 1.884 | 5.5 |
| 1.230 | 1.109 | 0.9 | 3.981 | 1.995 | 6.0 |
| 1.259 | 1.122 | 1.0 | 4.467 | 2.113 | 6.5 |
| (continued next page) | | | | | |

| Power Ratio | Voltage or Current Ratio | dB | Power Ratio | Voltage or Current Ratio | dB |
|---|---|---|---|---|---|
| 5.012 | 2.239 | 7.0 | 1 000 | 31.62 | 30 |
| 5.623 | 2.371 | 7.5 | 1 259 | 35.48 | 31 |
| 6.310 | 2.512 | 8.0 | 1 585 | 39.81 | 32 |
| 7.079 | 2.661 | 8.5 | 1 995 | 44.67 | 33 |
| 8.913 | 2.985 | 9.5 | 2 512 | 50.12 | 34 |
| 10.0 | 3.162 | 10 | 3 162 | 56.23 | 35 |
| 12.6 | 3.55 | 11 | 3 981 | 63.10 | 36 |
| 15.9 | 3.98 | 12 | 5 012 | 70.79 | 37 |
| 20.0 | 4.47 | 13 | 6 310 | 79.43 | 38 |
| 25.1 | 5.01 | 14 | 7 943 | 89.13 | 39 |
| 31.6 | 5.62 | 15 | 10 000 | 100.0 | 40 |
| 39.8 | 6.31 | 16 | 12 590 | 112.2 | 41 |
| 50.1 | 7.08 | 17 | 15 850 | 125.9 | 42 |
| 63.1 | 7.94 | 18 | 19 950 | 141.3 | 43 |
| 79.4 | 8.91 | 19 | 25 120 | 158.5 | 44 |
| 100.0 | 10.00 | 20 | 31 620 | 177.8 | 45 |
| 125.9 | 11.22 | 21 | 39 810 | 199.5 | 46 |
| 158.5 | 12.59 | 22 | 50 120 | 223.9 | 47 |
| 199.5 | 14.13 | 23 | 63 100 | 251.2 | 48 |
| 251.2 | 15.85 | 24 | 79 430 | 281.8 | 49 |
| 316.2 | 17.78 | 25 | 100 000 | 316.2 | 50 |
| 398.1 | 19.95 | 26 | 1 000 000 | 1,000 | 60 |
| 501.2 | 22.39 | 27 | 10 000 000 | 3,162 | 70 |
| 631.0 | 25.12 | 28 | 100 000 000 | 10,000 | 80 |
| 794.3 | 28.18 | 29 | 1 000 000 000 | 31,620 | 90 |

# 62. The Neper

The neper is also a dimensionless unit. The neper is used to express power ratios, but using the natural logarithm or logarithms to the base e (where e = 2.71828).

The following formula is used to calculate power ratios in nepers:

**Formula 62.1**

$$Np = 0.5 \times \ln \frac{Pout}{Pin}$$

Where:

Neper is the power ratio in nepers (n)

ln is the natural logarithm ($\log_n$)

Pout is the output power in watts (W)

Pin is the input power in watts (W)

**TABLE 22**
**Relationships Between Decibels and Nepers**

| |
|---|
| 1 decibel = 0.1 Bel |
| 1 decibel = 0.1151 neper |
| 1 bel = 1.151 nepers |
| 1 bel = 10 decibels |
| 1 neper = 0.8686 bel |
| 1 neper = 8.686 decibels |

# 63. BALANCED T-ATTENUATOR

The next formulas are used for calculations involving the symmetric-T attenuator.

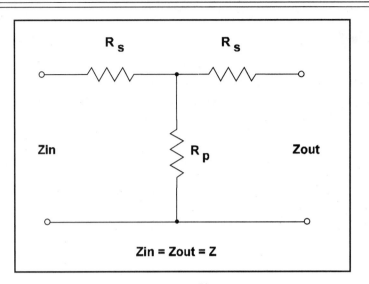

*Figure 52*

## Formula 63.1

$$Rs = Z \times \left( \frac{k-1}{k+1} \right)$$

Where:

Rs is the parallel resistance in ohms ($\Omega$)

Z is the impedance (Zin and Zout must be equal) in ohms ($\Omega$)

k is the attenuation factor (ratio of input voltage/current and output voltage/current)

## Formula 63.2

$$Rp = Z \times \left( \frac{2xk}{k^2 - 1} \right)$$

Where:

Rp is the parallel resistance in ohms ($\Omega$)

Z is the impedance (Zin = Zout) in ohms ($\Omega$)

k is the attenuation factor

**Application Example:**

Determine the resistances to be used in a T-attenuator where the input impedance and output impedance are 300 ohms and the desired attenuation factor is 4.

Data:

$Z = 300$ ohms

$k = 4$

$Rp = ?$

$Rs = ?$

Applying Formula 63.1 for Rs:

$$Rs = 300 \text{ x} \left( \frac{4-1}{4+1} \right) = 300 \text{ x } \frac{3}{5} = 180 \text{ ohms}$$

Applying Formula 63.2 to determine Rp:

$$Rp = 300 \text{ x} \left( \frac{2x4}{16-1} \right) = 300 \text{ x } \frac{8}{15} = 160 \text{ ohms}$$

# 64. BALANCED Pi-ATTENUATOR

The next formulas are used to calculate the com ponents of a Pi (π) attenuator.

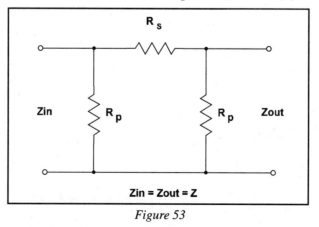

Zin = Zout = Z

*Figure 53*

## Formula 64.1

$$Rs = Z \times \left( \frac{k^2 - 1}{2xk} \right)$$

Where:

Rs is the series resistance in ohms ($\Omega$)

Z is the impedance (Zin=Zout) in ohms ($\Omega$)

k is the attenuation factor

## Formula 64.2

$$Rp = Z \times \left( \frac{k+1}{k-1} \right)$$

Where:

Rp is the parallel resistance in ohms ($\Omega$)

Z is the input/output impedance ($\Omega$)

k is the attenuation factor

**Application Example:**

Calculate the components of a Pi-attenuator with an input/output impedance of 50 ohms and an attenuation factor of 5.

Data:

Z = 50 ohms

k = 5

Rp, Rn = ?

Applying Formula 64.1 for Rs:

$$Rs = 50 \times \left( \frac{5^2 - 1}{2x5} \right) = 50 \times \left( \frac{24}{10} \right) = 50 \times 2.4 = 120 \text{ ohms}$$

Using Formula 64.2 for Rp:

$$Rp = 50 \times \left(\frac{5+1}{5-1}\right) = 50 \times \frac{6}{4} = 75 \text{ ohms}$$

# 65. UNBALANCED T-ATTENUATOR

In the unbalanced attenuator the output impedance is lower than the input impedance. The next formulas are used to determine the components in this attenuator:

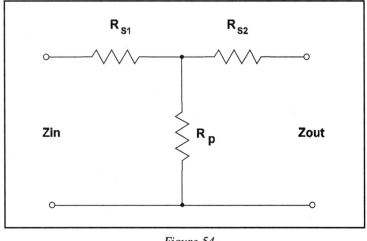

*Figure 54*

## Formula 65.1

Calculating the parallel resistance (Rp) in ohms ($\Omega$ )

$$Rp = \frac{2x\sqrt{NxZoutxZin}}{N-1}$$

Where:

N is attenuation factor (ratio between input/output voltages, currents or powers)

Zout is the output impedance in ohms ($\Omega$)

Zin is the input impedance in ohms ($\Omega$)

## Formula 65.2

$$Rs1 = Zin \times \left(\frac{N+1}{N-1}\right) - Rp$$

Where:

Rs1 is the series resistance in ohms ($\Omega$)

Zin is the input resistance in ohms ($\Omega$)

N is the attenuation factor

Rp is the parallel resistance in ohms ($\Omega$)

## Formula 65.3

$$Rs2 = Zout \times \left(\frac{N+1}{N-1}\right) - Rp$$

Where:

Rs2 is the series resistance in ohms ($\Omega$)

Zout is the output impedance in ohms ($\Omega$)

N is the attenuation factor

Rp is the parallel resistance in ohms ($\Omega$)

# 66. UNBALANCED Pi-ATTENUATOR

In this kind of attenuator the input impedance is always lower than the output impedance. The following formulas are used to calculate elements in this attenuator.

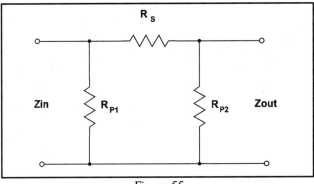

*Figure 55*

## Formula 66.1

Used to calculate the series resistance:

$$Rs = \frac{N-1}{2} \ \textbf{X} \ \sqrt{\frac{ZinxZout}{}N}$$

Where:

   Rs is the series resistance in ohms ($\Omega$)

   N is the attenuation factor

   Zin is the input impedance in ohms ($\Omega$)

   Zout is the output impedance in ohms ($\Omega$)

## Formula 66.2

Used to calculate Rp1 (resistance in parallel):

$$\frac{1}{Rp1} = \frac{1}{Zin} \ \textbf{X} \ \left(\frac{N+1}{N-1}\right) - \frac{1}{Rs}$$

Where:

   Rp1 is the parallel resistance in ohms ($\Omega$)

   Zin is the input impedance in ohms($\Omega$)

   N is the attenuation factor

   Rs is the series resistance ($\Omega$)

## Formula 66.3

$$\frac{1}{Rp2} = \frac{1}{Zout} \ \textbf{X} \ \left(\frac{N+1}{N-1}\right) - \frac{1}{Rs}$$

Where:

   Rp2 is the parallel resistance in ohms ($\Omega$)

   Zout is the output impedance in ohms ($\Omega$)

   N is the attenuation factor

   Rs is the series resistance in ohms

# 67. HALF-WAVE DIPOLE

The formulas given in this section are used to calculate the total length of a half-wave dipole as a function of the operation frequency.

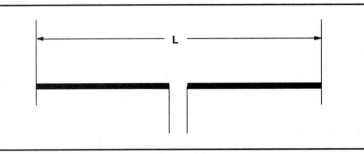

*Figure 56*

## Formula 67.1

$$L = \frac{\lambda}{2}$$

Where:

L is the antenna's length in meters (m)

$\lambda$ is the wavelength in meters (m)

## Formula 67.2

$$L = \frac{150000000}{f}$$

Where:

L is the antenna's length in meters (m)

f is the frequency in hertz (Hz)

**Application Example:**

Calculate the total length of a half-wave dipole cut to operate in a frequency of 30 MHz.

Data:

$\quad$ f = 30 MHz = 30 000 000

$\quad$ L = ?

Applying Formula 67.2

$$L = \frac{150000000}{30000000} = 5 \text{ meters}$$

# 68. FOLDED HALF-WAVE DIPOLE

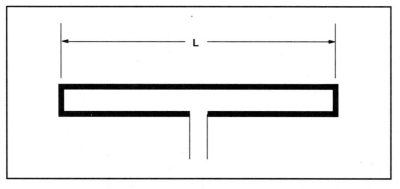

*Figure 57*

## Formula 68.1

$$L = \frac{\lambda}{2}$$

Where:

$\quad$ L is the total length of the antenna in meters (m)

$\quad$ $\lambda$ is the wavelength in the operation frequency in meters (m)

**Application Example:**

Determine the total length of a folded half-wave dipole to operate in a frequency of 300 MHz.

Data:

f = 300 MHz = 300 000 000 Hz

L = ?

Calculating the wavelength:

$$\lambda = \frac{300000000}{300000000} = 1 \text{ meter}$$

Using Formula 68.1:

$$L = \frac{1}{2} = 0.5 \text{ meter}$$

## TABLE 23
## Frequency x Wavelength for Some Frequencies

| Frequency (MHz) | Wavelength (meters) | Frequency (MHz) | Wavelength (meters) | Frequency (MHz) | Wavelength (meters) |
|---|---|---|---|---|---|
| 1 | 300 | 20 | 15 | 200 | 1.5 |
| 2 | 150 | 25 | 12 | 250 | 1.2 |
| 3 | 100 | 30 | 10 | 300 | 1 |
| 4 | 75 | 40 | 7.5 | 400 | 0.75 |
| 5 | 60 | 50 | 6 | 500 | 0.6 |
| 6 | 50 | 60 | 5 | 600 | 0.5 |
| 7 | 42.85 | 70 | 4.285 | 700 | 0.428 |
| 8 | 37.5 | 80 | 3.75 | 800 | 0.375 |
| 9 | 33.33 | 90 | 3.333 | 900 | 0.333 |
| 10 | 30 | 100 | 3 | 1 000 | 0.3 |
| 15 | 20 | 150 | 2 | | |

# 69. RANGE

The range of a transmitter operating in the VHF and upper bands depends on the transmitter and receiver antenna's height, and also the earth's diameter. The next formula can be used for ranges up to some hundreds of kilometers.

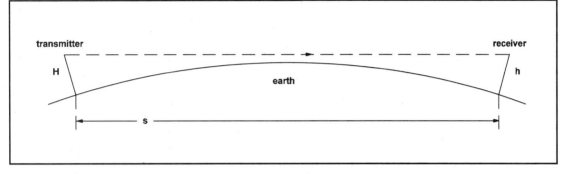

*Figure 58*

## Formula 69.1

$$S = 3.6 \times \left( \sqrt{H} \times \sqrt{h} \right)$$

Where:

S is the range in kilometers (km)

H is the transmitter's antenna height in meters (m)

h is the receiver's antenna height in meters (m)

NOTE: This formula assumes a region without hills, valleys or other obstacles.

**Application Example:**

What is the range of TV transmitter having a antenna height of 100 meters and received by an antenna placed 9 meters above the ground level?

Data:

H = 100 meters

h = 9 meters

S =?

Applying Formula 69.1:

$$S = 3.6 \times (\sqrt{100} + \sqrt{9}) = 3.6 \times (10 + 3) = 3.6 \times 13 = 46.8 \text{ kilometers}$$

# 70. COAXIAL CABLE

Impedance, capacitance and inductance of coaxial cable is calculated by the following formulas:

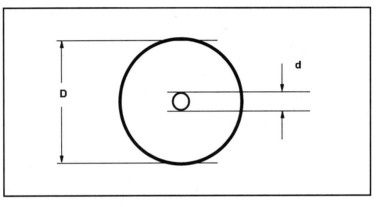

*Figure 59*

## Formula 70.1

$$Z = \frac{138}{\sqrt{\varepsilon'}} \times lg \, \frac{D}{d}$$

Where:

Z is the impedance in ohms ($\Omega$)

$\varepsilon$ is the dielectric constant (vacuum = 1)

D is the external diameter in centimeters (cm)

d is the internal diameter in centimeters (cm)

NOTE: Any unit can be used to express the diameters (internal and external), as the ratio between them is an absolute value.

### Formula 70.2

$$C = 0.24 \times \frac{\varepsilon x L}{\lg \dfrac{D}{d}}$$

Where:

C is the cable capacitance in pF

$\varepsilon$ is the dielectric constant

L is the length in centimeters (cm)

D is the external diameter in centimeters (cm)

d is the internal diameter in centimeters (cm)

### Formula 70.2

$$Lx = 0.046 \times \mu \times L \times \lg \frac{D}{d}$$

Where:

Lx is the inductance in microhenrys (uH)

$\mu$ is the relative permeability (for nonferromagnetic materials u = 1)

L is the length in centimeters (cm)

# 71. TWO-WIRE BALANCED LINE

The next formulas are used to determine the characteristics of 2-wire balanced lines.

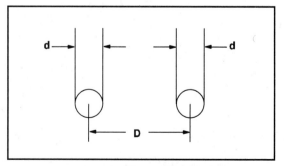

*Figure 60*

## Formula 71.1

$$Z = \frac{276}{\sqrt{\varepsilon_r}} \times \lg \frac{2D}{d}$$

Where:

Z is the impedance in ohms ($\Omega$)

$\varepsilon_r$ is the dielectric constant

D is the distance between wires in centimeters (cm)

d is the wire's diameter in centimeters (cm)

## Formula 71.2

$$C = 0.12 \times \frac{\varepsilon_{rxL}}{\lg \frac{2D}{d}}$$

Where:

C is the capacitance in picofarads (pF)

$\varepsilon_r$ is the dielectric constant

L is the length in centimeters (cm)

D is the distance between cables in centimeters (cm)

d is the diameter in centimeters (cm)

## Formula 71.3

$$Lx = 0.009 \times \mu \times L \times \lg \frac{2D}{d}$$

Where:

Lx is the inductance in microhenrys (uH)

$\mu$ is the relative permeability

L is the length in centimeters (cm)

D is the distance between wires in centimeters (cm)

d is the diameter in centimeters (cm)

# 72. IMPEDANCE MATCH - Pi-Network

The Pi-network can be used to match impedances; for instance, the ouput impedance of a transmitter with the impedance of an antenna.

The next formulas are used to calculate the filter's elements:

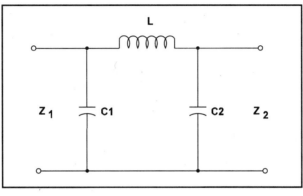

*Figure 61*

### Formula 72.1

$$\frac{Z1}{Z2} = \left(\frac{C2}{C1}\right)^2$$

Where:

Z1 is the imput impedance in ohms ($\Omega$)

Z2 is the output impedance in ohms ($\Omega$)

C1 and C2 are the capacitances in farads (F)

### Formula 72.2

$$Ct = \frac{C1xC2}{C1+C2}$$

Where:

Ct is the total capacitance in picofarads (pF)

C1 and C2 are the capacitances in the network in picofarads (pF)

## Formula 72.2

$$f = \frac{1}{2 \times \pi \times \sqrt{L \times C_t}}$$

Where:

f is the resonant frequency in hertz (Hz)

$\pi$ is the constant 3.1416

L is the associated inductance in henrys (H)

Ct is the equivalent capacitance in farads (F)

NOTE: You can use as units the microhenry and the picofarad to find the frequency in megahertz.

## Formulas 73.3

$$Z1 = Z2 \times \left(\frac{C2}{C1}\right)^2$$

$$Z2 = Z1 \times \left(\frac{C1}{C2}\right)^2$$

Where:

Z1 is the input impedance in ohms ($\Omega$)

Z2 is the output impedance in ohms ($\Omega$)

C1, C2 are the capacitances in farads (F)

# Part 3

# Electronic Circuits

# 73. SEMICONDUCTOR DIODE

To an ideal semiconductor diode (p-n junction), the current flowing when direct-biased depends on the voltage and temperature, as in the following formula:

**Formula 73.1**
**Diode current**

$$I = Irmax = \left[ e^{\left(\frac{U}{UT}\right)} - 1 \right]$$

Where:

I is direct current in amperes (A)

Irmax is the blocking current or saturation current in amperes (A)

U is applied voltage in volts (V)

Ur is the termic potential in volts (V)

e = 2.71828

# 74. HALF-WAVE RECTIFIER

This rectifier uses one diode and furnishes a DC pulsed voltage with the same frequency of the AC input. The following formulas are used to calculate the current and voltage across a load.

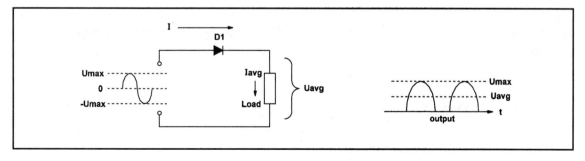

*Figure 62*

## Formula 74.1
## Average Voltage

$$\text{Uavg} = \frac{Umax}{\pi}$$

Where:

Uavg is the average voltage across the load in volts (V)

Umax is the voltage peak or maximum in volts (V)

$\pi$ is the constant 3.1416

NOTE: The voltage falls in the diodes (about 0.2V in germanium types and 0.6V in silicon types) were not considered by this formula.

## Formula 74.2
## Average Current

$$\text{Iavg} = \frac{Imax}{\pi}$$

Where:

Iavg is the average current through the load in amperes (A)

Imax is the peak or maximum AC input current in amperes (A)

$\pi$ is 3.1416

**Application Example:**

Determine the average current in a half-wave power supply where the maximum rectified current is 600 mA.

Data:

Imax = 600 mA

$\pi$ = 3.14

Iavg = ?

Using Formula 74.2:

$$Iavg = \frac{600}{3.14} = 191.08 \text{ mA}$$

NOTE: You can also use Imax in amperes, calculating the Iavg in amps.

# 75. FULL-WAVE RECTIFIER

Current and voltage in a full-wave rectifier can be calculated by the next formulas. The formulas are valid to 2-diode systems and the 4-diode bridge rectifier.

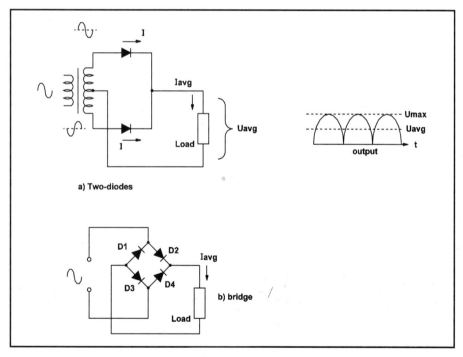

*Figure 63*

## Formula 75.1:
## Average Voltage

$$Uavg = \frac{2 \times Umax}{\pi}$$

Where:

Uavg is the medium voltage across the load in volts (V)

Umax is the peak voltage or maximum voltage in volts (V)

$\pi$ is 3.1416

## Formula 75.2
## Average Current

$$Iavg = \frac{2x\,Imax}{\pi}$$

Where:

Iavg is the average current through the load measured in amperes (A)

Imax is the peak current or maximum current in amperes (A)

$\pi$ is 3.1416

## Derivated Formula:

## Formula 75.3
## Maximum Current

$$Imax = \frac{Iavgx\,\pi}{2}$$

Where:

Imax is the maximum current in amperes (A)

Iavg is the average current in amperes (A)

$\pi$ is 3.1416

NOTE: The voltage fall in the diodes is not considered by this calculation.

**Application Example:**

Calculate peak voltage in a full-wave rectifier where the average output voltage in a load is 9V.

Data:

Uavg = 9 V

$\pi$ = 3.14

Umax = ?

Applying formula:

$$\text{Umax} = \frac{9 \times 3.14}{2} = 14.13 \text{ volts}$$

# 76. LC FILTER COEFFICIENT

The filter coefficient is defined as the ratio between the amplitude of the alternating component at its input and the amplitude at the output. The filter coefficient is calculated by the following formulas:

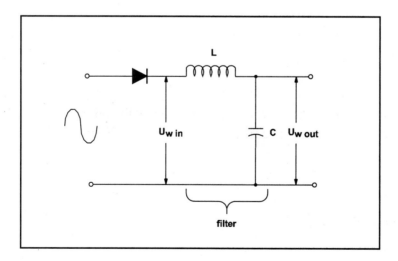

*Figure 64*

## Formula 76.1
## Filter Coefficient I

$$\alpha = \frac{Uwin}{Uwout}$$

Where:

$\alpha$ is the filter coefficient

Uwin is alternating component at the filter's input in volts (V)

Uwout is alternating component at the filter's output in volts (V)

## Formula 76.2
## Filter Coefficient II

$$\alpha = \omega^2 \, C \, L - 1$$

Where:

$\alpha$ is the filter coefficient

$\omega = (2 \times \pi \times f)$, where f is the frequency of the AC at the filter's input (Hz)

C is the capacitance in farads (F)

L is the inductance in henrys (H)

**Application Example:**

Determine the amplitude of the alternating component at the output of an RC filter operating at 60 Hz with an output of 12V. The filter is formed by a 1000 uF capacitor and a 10H filter:

Data:

L = 10 H

C = 1 000 uF = $10^{-3}$ F

f = 60 Hz

Uwin = 12V

Uwout = ?

Applying Formula 76.2 to calculate $\alpha$ :

$$\alpha = [ (2 \times 3.14 \times 60)^2 \times 10^{-3} \times 10) - 1] = 141.9 - 1 = 140.9$$

Calculating Uwout:

$$\text{Uwout} = \frac{12}{140.9} = 0.085V \text{ or } 85 \text{ mV}$$

# 77. RC FILTER COEFFICIENT

The filter coefficient to an RC configuration is defined as the ratio between Uwout and Uwin as in the LC filter. The next formulas are used to calculate this coefficient:

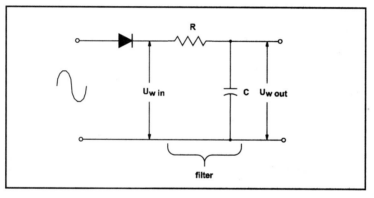

*Figure 65*

**Formula 77.1**
**Filter Coefficient**

$$\alpha = \frac{Uwout}{Uwin}$$

Where:

$\alpha$ is the filter coefficient

Uwout is the amplitude of the alternating component at filter's output in volts (V)

Uin is the amplitude of the alternating component at the filter's input in volts (V)

**Formula 77.2**

**Filter Coefficient as Function of RC**

$$\alpha = \sqrt{(\omega x R x C)^2 + 1}$$

Where:

$\alpha$ is the filter coefficient

$\omega = 2 \times \pi \times f$, where f is the VAC frequency at the filter's input in hertz (Hz)

R is the resistance in ohms ($\Omega$)

F is the capacitance in farads (F)

# 78. RIPPLE FACTOR

The ripple factor ($\gamma$) is defined as the ratio between the rms value of the output voltage and the DC value of the output voltage, times 100. The next formula expresses this definition and is used to calculate the ripple factor:

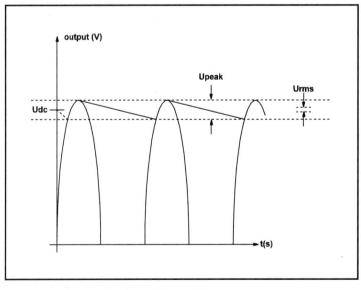

*Figure 66*

# Formula 78.1
## Ripple Factor

$$\gamma = \frac{Urms}{Udc} \text{ x } 100$$

Where:

$\gamma$ is the ripple factor

Urms is the rms value of the output voltage in volts (V)

Udc is the average value of the output voltage in volts (V)

**Ripple factors (Resistive load):**

Half-wave rectifier = 120%

Full-wave rectifier = 48%

# TABLE 24
## Rectifier's Characteristics (using resistive loads)

| Parameter | Half-Wave | Full-Wave(Cntr-tap transformer) | Full-Wave (Bridge) |
|---|---|---|---|
| Udc in the load | $\dfrac{Umax}{\pi}$ | $\dfrac{2xUmax}{\pi}$ | $\dfrac{2xUmax}{\pi}$ |
| Urms in the load | $\dfrac{Umax}{2}$ | $\dfrac{Umax}{\sqrt{2}}$ | $\dfrac{Umax}{\sqrt{2}}$ |
| Reverse voltage across the diodes Ur = (max) | Umax | 2 x Umax | Umax |
| Ripple factor ($\gamma$) | 120% | 48% | 48% |
| Transformer: storage factor referred to the output power in the load | 3.49 x Pdc (primary & secondary) | 1.75 x Pdc (secondary) 1.23 x Pdc (primary) | 1.23 x Pdc (primary and secondary) |

Pdc = output power in watts (W)

# 79. FILTER INDUCTANCE

The next formula is used to calculate the inductance to be used in a filter. This values depends on the filter coeffcient, the capacitor used, the output current and also the frequency of the AC applied to the filter.

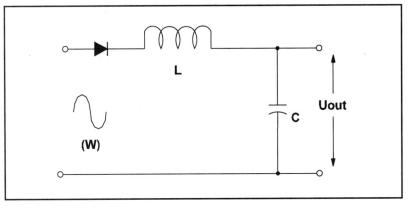

*Figure 67*

## Formula 79.1
## Filter Inductance

$$L = \frac{1+\alpha}{\omega^2 x C}$$

Where:

L is the inductance in henrys (H)

$\alpha$ is the filter coefficient

$\omega = 2 \times \Pi \times f$ , where f is the frequency in hertz (Hz)

C is the capacitance in farads (F)

NOTE: We can define the quantity $\eta$ as the filter efficiency and the value is given by the following formula:

## Formula 79.1a

## Efficiency

$$\eta = \frac{1}{\alpha}$$

Where:

$\alpha$ is the filter coefficient

$\eta$ is the filter effcience

# 80. FILTER CAPACITANCE

The filter capacitor can be calculated as a function of the filter inductance, the AC frequency and the filter coefficient. The next formula is used to make these calculations (see *Figure 67*):

## Formula 80.1
## Filter Capacitance:

$$C = \frac{1+\alpha}{\omega^2 x L}$$

Where:

C is the filter capacitance in farads (F)

$\alpha$ is the filter coefficient

$\omega = 2 \times \pi \times f$, where f is the AC frequency in hertz (Hz)

L is the coil inductance in henry (H)

## Derivated Formula:

## Formula 80.2
## Filter Capacitance II

$$C = \frac{1+\eta}{\eta x \omega^2 x L}$$

Where all the quantities are as in the previous formula, except $\eta$ is the filter efficiency

# 81. CONVENTIONAL VOLTAGE DOUBLER

*Figure 68* shows the basic configuration for a conventional voltage doubler. The working voltage of the capacitor (minimum WVDC) is calculated by the next formula:

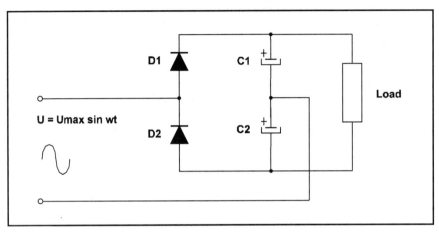

*Figure 68*

**Formula 81.2**
**Voltage Doubler**

$$Uc1 = Uc2 = Umax$$

Where:

Uc1 and U2 are the capacitors' working voltage in volts (V)

Umax is the peak value of the input AC voltage in volts (V)

NOTE: Minimum PIV of the diodes is 2 x Umax.

# 82. CASCADE VOLTAGE DOUBLER

This circuit uses only two diodes. The capacitor's working voltage is calculated as follows:

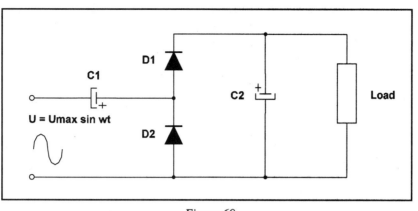

*Figure 69*

**Formulas 82.1**
**Cascade Voltage Doubler**

$$Uc1 = Umax$$

$$Uc2 = 2 \times Umax$$

Where:

Uc1 and Uc2 are the capacitors' working voltage in volts (V)

Umax is the peak or maximum value of the input AC voltage in volts (V)

NOTE: PIV of the diodes are 2 x Umax (minimum).

# 83. BRIDGE VOLTAGE DOUBLER

*Figure 70* shows the configuration using 4 diodes where the output voltage is two times the input voltage.

**Formula 83.1**
**Voltage in the Capacitors**

$$Uc1 = Uc2 = Umax$$

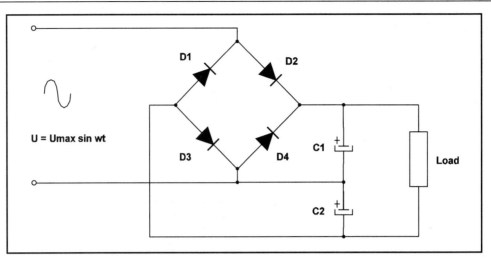

*Figure 70*

Where:

Uc1 and Uc2 are the minimum working VDC (WVDC) of the capacitors in volts (V)

Umax is the peak value of the applied input voltage in volts (V)

NOTE: Mininum PIV of the diodes is 2 x Umax.

# 84. FULL WAVE TRIPLER

The circuit shown in *Figure 71* furnishes in the output a voltage threc times greater than the input voltage. Components are calculated as follows.

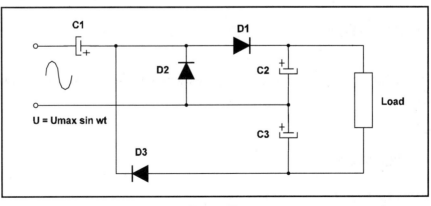

*Figure 71*

## Formulas 84.1
## Voltage in the Capacitors

$$Uc1 = Uc3 = Umax$$

$$Uc2 = 2 \times Umax$$

Where:

Uc1, Uc2, Uc3 are the working minumum voltage DC of the capacitors in volts (V)

Umax is the peak of the input AC voltage

NOTE: Minimum PIV of the diodes is 2 x Umax.

# 85. CASCADE VOLTAGE TRIPLER

The circuit given in *Figure 72* has a voltage output three times greater than the input voltage. The capacitor's characteristics are calculated as follows:

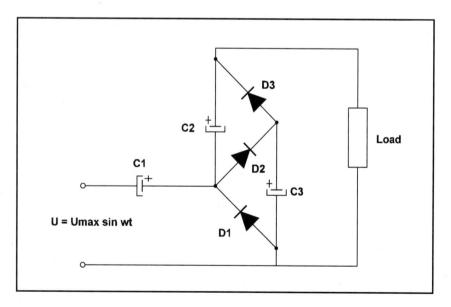

*Figure 72*

## Formulas 85.1
### Voltage in the Capacitors

$$Uc1 = Umax$$

$$Uc2 = Uc3 = 2 \times Umax$$

Where:

Uc1, Uc2, Uc3 are the minimum working VDC (WVDC) of the capacitors in volts (V)

Umax is the peak of the input AC voltage in volts (V)

NOTE: Minimum PIV of the diodes is 2 x Umax.

# 86. FULL-WAVE VOLTAGE QUADRUPLER

*Figure 73* shows the circuit using 4 diodes. The minimum working voltage DC of the used capacitors are calculated as follows:

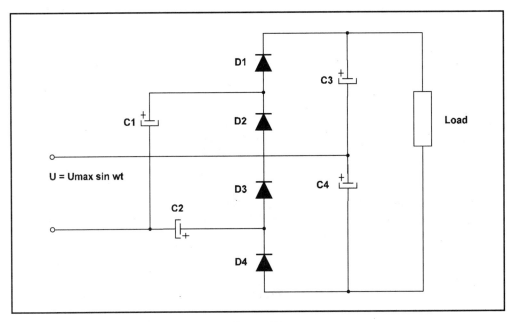

*Figure 73*

## Formula 86.1
### Voltage in the Capacitors

$$Uc1 = Uc2 = Uc3 = Uc4 = 2 \times Umax$$

Where:

Uc1, Uc2, Uc3, Uc4 are the minimum working voltage DC (WVDC) of the used capacitors in volts (V)

Umax is the peak value of the input AC voltage in volts (V)

NOTE: Minimum PIV of the used diodes is 2 x Umax.

# 87. ZENER DIODE

The following formulas are used in calculations related to a zener voltage regulator circuit, such as the one shown in *Figure 74*.

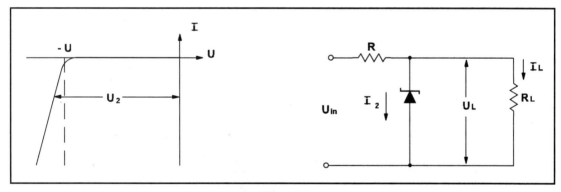

*Figure 74*

## Formulas 87.1
### Maximum Current through a Zener Diode

$$Iz(max) = \frac{Uin(max) - (U_L + RxI_L)}{R}$$

$$Iz(min) = \frac{Uin(min) - (U_L + RxI_L)}{R}$$

Where:

Iz(max) is the maximum current through the zener in amperes (A)

Iz(min) is the minimum current through the zener in amperes (A)

Uin(min) is the minimum input voltage in volts (V)

Uin(max) is the maximum input voltage in volts (V)

$U_L$ is the output voltage or load voltage in volts (V)

$I_L$ is the output current in amperes (A)

R is the load resistance in ohms ($\Omega$ )

## Formulas 87.2

The next formulas are used to calculate the range of values of R.

$$R(min) = \frac{Uin(max) - U_L}{Iz(max) + I_L}$$

$$R(max) = \frac{Uin(min) - U_L}{Iz(min) + I_L}$$

Where:

R(min) is the minimum value of R in ohms ($\Omega$)

R(max) is the maximum value of R in ohms ($\Omega$)

Uin(max) is the maximum input voltage in volts (V)

Uin(min) is the minimum input voltage in volts (V)

Iz(min) is the minimum zener current in amperes (A)

Iz(max) is the maximum zener current in amperes (A)

$U_L$ is the voltage across the load in volts (V)

$I_L$ is the load current in volts (V)

# Formula 87.3
# Maximum Zener Current

$$Iz(max) = \frac{P(max)}{Uz}$$

Where:

Iz(max) is the maximum current in the zener diode in amperes (A)

P(max) is the maximum zener's power dissipation in watts (W)

Uz is the zener voltage in volts (V)

# Derivated Formula:

# Formula 87.4
# Maximum Zener Dissipation

$$P(max) = Uz \times Iz(max)$$

Where:

P(max) is the maximum power dissipated by the zener in watts (W)

Uz is the zener voltage

Iz(max) is the maximum current through the zener in amperes (A)

**Application Example:**

A 20 mA load must be powered from a 9V regulated supply. The input voltage to the regulator circuit can change its value between 12 and 15 volts, and the current through the zener must be kept between 10 and 50 mA. Calculate the value and dissipation of R.

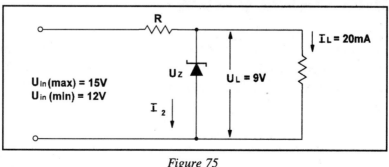

*Figure 75*

Data:

$I_L = 20$ mA

$U_L = 9V$

Uin(max) = 15V

Uin(min) = 12V

Iz(max) = 50 mA = 0.05A

Iz(min) = 10 mA = 0.01A

Calculating R(min) and R(max), using Formula 87.2:

$$R(min) = \frac{15-9}{0.05-0.02} = \frac{6}{0.03} = 200 \text{ ohms}$$

$$R(max) = \frac{12-9}{0.02-0.01} = \frac{3}{0.01} = 300 \text{ ohms}$$

(We can adopt 220 ohms, as it is a standard value for resistors and is in the calculated range.)

Determining the dissipated power, using formula 87.4:

$$P(max) = 0.05 \times 9 = 0.45 \text{ Watt}$$

(A 1W zener is recommended for the application.)

# 88. CAPACITIVE VOLTAGE DIVIDER

The capacitive reactance of a capacitor can be used in a voltage divider, from which we can design a transformerless power supply, such as the one shown in *Figure 76*. Supposing a resistive load we can calculate C and R using the next formulas:

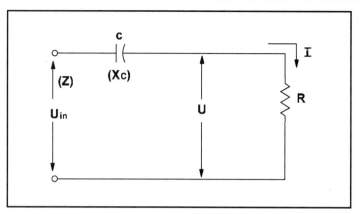

*Figure 76*

## Formula 88.1
## Calculating R

$$R = \frac{U}{I}$$

Where:

U is voltage across the load in volts (V)

I is the current in the load in amperes (A)

R is the load resistance in ohms ($\Omega$)

NOTE: U and I are root-mean-square (rms) values.

## Formula 88.2
## Impedance

$$Z = \frac{Uin}{I}$$

Where:

Z is the circuit impedance in ohms ($\Omega$)

Uin is the input voltage in volts (V)

I is the total circuit current or load current in amperes (A)

## Formula 88.3
## Capacitive Reactance

$$Xc = \sqrt{Z^2 - R^2}$$

Where:

Xc is the capacitive reactance of the used capacitor in farads (F)

Z is the circuit impedance in ohms ($\Omega$ )

R is the load resistance in ohms ($\Omega$)

## Formula 88.4

$$C = \frac{10^6}{2 \times \pi \times f \times Xc}$$

Where:

C is the capacitance in microfarads ($\mu$ F)

$\pi$ is the constant 3.1416

f is the AC power line frequency in hertz (Hz)

Xc is the capacitive reactance in ohms ($\Omega$ )  .

**Application Example:**

Determine the capacitance of C in the circuit shown in *Figure 77* to power the 12V x 50 mA lamp from the 117V x 60 Hz power line.

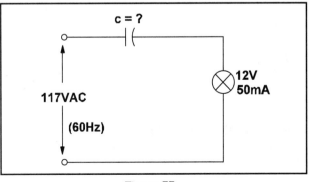

*Figure 77*

Data:

U = 12V

Uin = 117V

I = 0.02A (20 mA)

f = 60 Hz

C = ?

Calculating R, using Formula 88.1:

$$R = \frac{12}{0.02} = 600 \text{ ohms}$$

Determining Z, using Formula 88.2:

$$Z = \frac{117}{0.02} = 5\,850 \text{ ohms}$$

Now, Xc can be found, using Formula 88.3:

$$Xc = \sqrt{(5{,}850)^2 - (600)^2} = \sqrt{31.9x10^6 - 0.36x10^6} = \sqrt{31.56x10^6} = 5618 \text{ ohms}$$

Finding C, using Formula 88.4:

$$C = \frac{10^6}{2x3.14x60x5{,}618} = \frac{10^6}{2.116x10^6} = \frac{1}{2.116} = 0.47 \ \mu F \text{ or } 470 \text{ nF}$$

# 89. NTC

Negative Temperature Coefficient resistors, or NTCs, are devices having a characteristic as shown in *Figure 78*. The resistance decreases when the temperature rises.

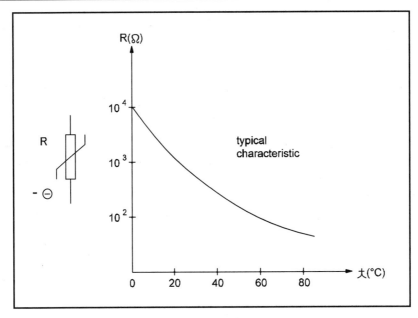

*Figure 78*

## Formula 89.1
## Resistance of an NTC

$$R = A \times e^{\frac{B}{T}}$$

Where:

R is the resistance in a given temperature in °K in ohms ($\Omega$)

A is a constant given by the manufacturer of the NTC.

B is a constant given by the characteristic (range from 1,000 to 8,000 °K)

T is the ambient temperature in degrees Kelvin (°K)

## Formula 89.2
## Resistance of an NTC II

$$R = R_{to} \times e^{\left[ B\left( \frac{1}{T} - \frac{1}{T_o} \right) \right]}$$

Where:

R is the resistance in a given temperature in ohms ($\Omega$)

Rto is the resistance at To in ohms ($\Omega$)

e = 2.71828

T is the final temperature in degrees kelvin (°K)

To is the the reference temperature in degress kelvin (°K)

B is a constant given by the component characteristic

# 90. PTC

Positive Temperature Coefficient resistors are components having a characteristic as shown in *Figure 79*. Their resistance increases as the temperature rises.

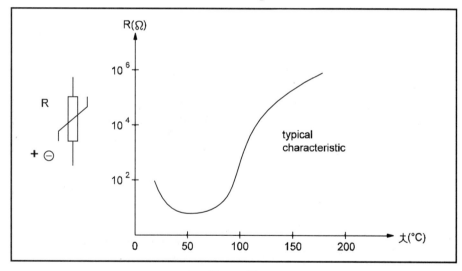

*Figure 79*

## Formula 90.1
## Resistance

$$\alpha = \frac{\ln\left(\dfrac{R2}{R1}\right)}{\Delta t}$$

Where:

$\alpha$ is the coefficient temperature above the reference temperature in degrees celsius.

R1 and R2 are resistances at two different points of the characteristic curve above the reference temperature in ohms ($\Omega$)

$\Delta$ t is the temperature difference between the points where R1 and R2 are measured.

NOTE: Characteristic curves, given by the PTC manufacturer, depend on each device.

# 91. VARICAPS

Varicaps, varactors or variable-capacitance diodes, are semiconductor devices whose capacitance depends on the reverse-bias voltage. These components are used in tuned circuits to change the frequency from an external control voltage.

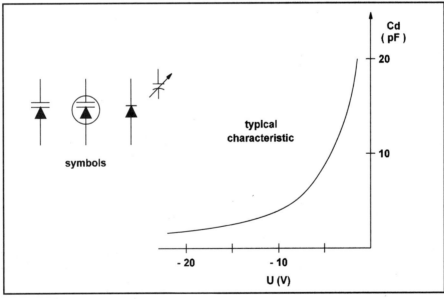

*Figure 80*

**Formula 91.1**

**Varicap**

$$Cd = Co \times \left( \frac{Uo}{U + Uo} \right)^{\frac{1}{n}}$$

Where:

Cd is the final capacitance in picofarads (pF)

Co is the capacitance when U = 0V

n is the index given by the manufacturer according to the device

U is the applied voltage in volts (V)

Uo is the diffusion voltage in volts (V)

NOTE: Uo = 0.5V (typ) for germanium
  Uo = 0.7V (typ) for silicon

# TRANSISTORS

## 92. TRANSISTOR STATIC CURRENT-GAIN (Common Emiter)

The static current gain for a transistor is the ratio between the colector's continuous current (Ic) and the base's continuous current (Ib) in a common-emitter configuration, as shown in *Figure 81*.

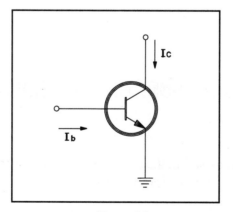

*Figure 81*

**Formula 92.1**
**Transistor Gain**

$$\beta = \frac{Ic}{Ib}$$

Where:

$\beta$ is the static current gain

Ic is the collector current in amperes (A) or submultiples

Ib is the base current in amperes (A) or submultiples

**Application Example:**

A current change of 1 mA in the base of a transistor produces a colector current change of 50 mA. The transistor is in a common-collector configuration. Calculate the static gain of this transistor.

Data:

Ic = 50 mA

Ib = 1 mA

$\beta = ?$

Using Formula 92.1:

$$\beta = \frac{50}{1} = 50$$

# 93. TRANSISTOR STATIC CURRENT GAIN (Common-Base Configuration)

The static current gain of a transistor in the common-base configuration is defined as the ratio between emitter current (Ie) and base current (Ib).

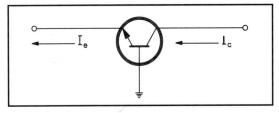

*Figure 82*

## Formula 91.1
## Current Gain

$$\alpha = \frac{Ie}{Ib}$$

Where:

$\alpha$ is the current gain in common base configuration

Ie is the emitter current in amperes (A)

Ib is the base current in amperes (A)

# 94. RELATIONSHIP BETWEEN ALPHA AND BETA

The following formula is used to convert alpha into beta and vice-versa.

## Formulas 94.1
## Alfa vs. Beta

$$\beta = \frac{\alpha}{1-\alpha}$$

$$\alpha = 1 - \frac{1}{\beta - 1}$$

Where:

$\beta$ is the static current gain in the common-collector configuration

$\alpha$ is the static current gain in the common-base configuration

**Application Example:**

Determine the static current gain in common-emitter configuration ($\beta$) of a transistor with a static current gain in common-base configuration ($\alpha$) of 0.995.

Data:

$$\alpha = 0.995$$

$$\beta = ?$$

Using Formula 94.1:

$$\beta = \frac{0.995}{1 - 0.995} = \frac{0.995}{0.005} = 199$$

## TABLE 25
## Alpha & Beta Conversion

| $\beta$ (Beta) | $\alpha$ (Alpha) | $\beta$ (Beta) | $\alpha$ (Alpha) |
|:---:|:---:|:---:|:---:|
| 1 | 0.5000 | 16 | 0.9412 |
| 2 | 0.6666 | 17 | 0.9444 |
| 3 | 0.7500 | 18 | 0.9474 |
| 4 | 0.8000 | 19 | 0.9500 |
| 5 | 0.8333 | 20 | 0.9524 |
| 6 | 0.8571 | 22 | 0.9565 |
| 7 | 0.8750 | 24 | 0.9600 |
| 8 | 0.8889 | 26 | 0.9630 |
| 9 | 0.9000 | 28 | 0.9655 |
| 10 | 0.9091 | 30 | 0.9677 |
| 11 | 0.9167 | 32 | 0.9697 |
| 12 | 0.9231 | 34 | 0.9714 |
| 13 | 0.9286 | 36 | 0.9730 |
| 14 | 0.9333 | 38 | 0.9744 |
| 15 | 0.9375 | 40 | 0.9756 |

(continued next page)

| $\beta$ (Beta) | $\alpha$ (Alpha) | $\beta$ (Beta) | $\alpha$ (Alpha) |
|---|---|---|---|
| 42 | 0.9767 | 150 | 0.9933 |
| 44 | 0.9778 | 160 | 0.9938 |
| 46 | 0.9786 | 170 | 0.9942 |
| 48 | 0.9796 | 180 | 0.9945 |
| 50 | 0.9804 | 190 | 0.9948 |
| 55 | 0.9821 | 200 | 0.9952 |
| 60 | 0.9836 | 250 | 0.9962 |
| 65 | 0.9848 | 300 | 0.9966 |
| 70 | 0.9859 | 350 | 0.9971 |
| 75 | 0.9868 | 400 | 0.9975 |
| 80 | 0.9877 | 450 | 0.9978 |
| 85 | 0.9884 | 500 | 0.9980 |
| 90 | 0.9890 | 550 | 0.9982 |
| 95 | 0.9896 | 600 | 0.9983 |
| 100 | 0.9901 | 650 | 0.9984 |
| 110 | 0.9909 | 700 | 0.9986 |
| 120 | 0.9917 | 750 | 0.9987 |
| 130 | 0.9931 | 800 | 0.9987 |
| 140 | 0.9932 | | |

# 95. HYBRID PARAMETERS

The hybrid parameters were developed by transistor manufacturers to indicate the values of their internal elements. Using hybrid parameters, calculations involving transistor circuits can be made easier. *Figure 83* shows the the equivalent circuit for such hybrid parameters.

**Definitions:**

The hybrid parameters in the circuit of *Figure 83* are defined as follows:

- $h_{11}$ input impedance with short-circuit output
- $h_{12}$ reverse open-circuit voltage amplification factor
- $h_{21}$ forward short-circuit current amplification factor
- $h_{22}$ output admittance with open-circuit input

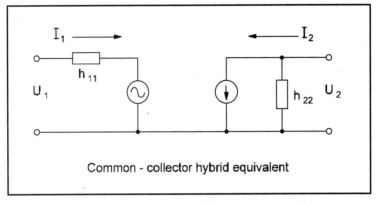

Common - collector hybrid equivalent

*Figure 83*

## Formula 95.1
## Input Impedance

$$h_{11} = \frac{U_1}{I_1} \quad \text{for} \quad U_2 = 0$$

## Formula 95.2
## Reverse Open-Circuit Voltage Amplification Factor

$$h_{12} = \frac{U_1}{U_2} \quad \text{for} \quad I_1 = 0$$

## Formula 95.3
## Forward Short-Circuit Current Amplification Factor ( $\beta$ )

$$h_{21} = \frac{I_2}{I_1} \quad \text{for} \quad U_2 = 0$$

## Formula 95.4
## Output Admittance

$$h_{22} = \frac{I_2}{U_2} \quad \text{for} \quad I_1 = 0$$

Where:

$U_1$ is the input voltage in volts (V)

$U_2$ is the output voltage in volts (V)

$I_1$ is the input current in amperes (A)

$I_2$ is the output current in amperes (A)

## Derivated Formulas:

## Formula 95.5
## Input Voltage

$$U_1 = h_{11} \times I_1 + h_{12} \times U_2$$

## Formula 95.6
## Output Current

$$I_2 = h_{21} \times I_1 + h_{22} \times U_2$$

## Formula 95.7
## Determinant h

$$\det h = h_{11} \times h_{22} - h_{12} \times h_{21}$$

Where: det h is the determinant formed by the hybrid parameters as follows:

$$\det h = \begin{vmatrix} h_{11} & h_{21} \\ h_{12} & h_{22} \end{vmatrix}$$

**TABLE 26**

**Letter Symbols and Abbreviations for Transistors**

(Applied to Formulas 95 to 105)

| Symbol | Term |
| --- | --- |
| Det h | Determinant h |
| Gi | Current gain |
| Gv | Voltage gain |
| Gp | Power gain |
| $h_{11}$ | Input impedance with short-circuit output |
| $h_{12}$ | Reverse open-circuit voltage amplification factor |
| $h_{21}$ | Forward short-circuit amplification factor |
| $h_{22}$ | Output admittance with open circuit output |
| $I_1$ | Input current |
| $I_2$ | Output current |
| Ib | Base current |
| Ic | Collector current |
| Ie | Emitter current |
| $R_G$ | Generator resistance |
| $R_L$ | Load resistance |
| $U_1$ | Input voltage |
| $U_2$ | Output voltage |
| $U_{bc}$ | Voltage between base and collector |
| $U_{be}$ | Voltage between base and emitter |
| $U_{ce}$ | Voltage between collector and emitter |

# 96. COMMON-BASE

*Figure 84* shows the common base configuration using an NPN transistor. In this configuration the signal to be amplified is applied to the emitter and taken from the collector. The base is the common element to the input and output signals as it is wired to ground.

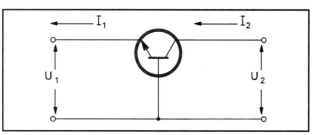

*Figure 84*

## Formulas 96.1
## Common Base

$$U_1 = U_{be}$$
$$I_1 = I_e$$
$$U_2 = U_{bc}$$
$$I_2 = I_c$$

Where:

$U_{be}$ is the voltage between base and emitter in volts (V)

$I_e$ is the current flowing through the emitter in amperes (A)

$U_{bc}$ is the voltage between base and collector in volts (V)

$I_c$ is the current flowing through the collector in amperes (A)

# 97. COMMON EMITTER

This configuration is shown in *Figure 85*. The signal is applied to the base and is taken from the collector. The emitter is the common element to the input and output. The next formulas are used when calculating the hybrid parameters related to this circuit:

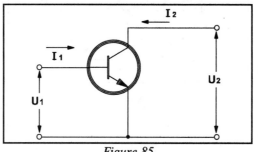

*Figure 85*

**Formulas 97.1**
**Common-Emitter**

$$U_1 = U_{eb}$$
$$I_1 = I_b$$
$$U_2 = U_{ce}$$
$$I_2 = I_c$$

Where:

$U_{eb}$ is the voltage between emitter and base in volts (V)

$I_b$ is the current flowing through the base in amperes (A)

$U_{ce}$ is the voltage between collector and emitter in volts (V)

$I_c$ is the current flowing through the collector in amperes (A)

# 98. COMMON-COLLECTOR

In the common-collector configuration the signal is applied to the base and taken from the emitter. The collector is the element common to input and output circuits. *Figure 86* shows this configuration. The next formulas are used to calculate the hybrid parameters.

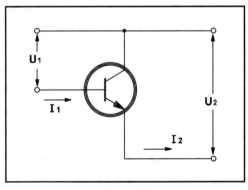

*Figure 86*

## Formulas 98.1
## Common-Collector

$$U_1 = U_{bc}$$
$$I_1 = I_b$$
$$U_2 = U_{ce}$$
$$I_2 = I_e$$

Where:

$U_{cb}$ is the voltage between base and collector in volts (V)

$I_b$ is the current flowing through the base in amperes (A)

$U_{ce}$ is the voltage between collector and emitter in volts (V)

$I_e$ is the current flowing through the emitter in amperes (A)

## TABLE 27
## General Characteristics of Transistor Circuits

| Common-Emitter | Common-Base | Common-Collector |
|---|---|---|
| Large voltage gain | Large voltage gain | Near unity voltage gain |
| Large current gain | Near unity current gain | Large current gain |
| Highest power gain | Medium power gain | Lowest power gain |
| Low input resistance | Very low input resistance | High input resistance |
| High output resistance | Very high output resistance | Low output resistance |

# BASIC QUANTITIES OF CIRCUITS USING TRANSISTORS

The following formulas are valid when making calculations involving the characteristic quantities of three basic transistor configurations (common-base, common-emitter and common-collector) applied to low-frequency and DC circuits. See *Figure 87*.

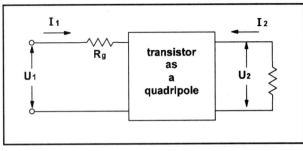

*Figure 87*

# 99. SHORT-CIRCUIT OUTPUT

$R_L = 0$ and $U_2 = 0$

The following formulas are used to calculate characteristic quantities in all configurations.

## Formula 99.1
## Input Resistance

$$\frac{U_1}{I_1} = h_{11}$$

Where the quantities are the same as defined in *Table 26*.

## Formula 99.2
## Input Conductance

$$\frac{I_1}{U_1} = \frac{1}{h_{11}}$$

Where the quanties are the same as defined in *Table 26*.

## Formula 99.3
## Current Gain

$$\frac{I_2}{I_1} = h_{21}$$

Where the quantities are the the same as defined in *Table 26*.

## Formula 99.4
### Transfer Conductance

$$\frac{I_2}{U_1} = \frac{h_{21}}{h_{11}}$$

Where the quantities are the the same as used in *Table 26*.

# 100. OPEN-CIRCUIT OUTPUT

$R_L = \infty$ and $I_2 = 0$

($R_L$ is the load resistance)

The following formulas are used to calculate the characteristic quantities when the circuit in any configuration operates with an open-circuit output.

## Formula 100.1
### Input Resistance

$$\frac{U_1}{I_1} = \frac{\det h}{h_{22}}$$

Where the quantities are as defined in *Table 26*.

## Formula 100.2
### Input Conductance

$$\frac{I_1}{U_1} = \frac{h_{22}}{\det h}$$

Where the quantities are as defined in *Table 26*.

**Formula 100.3**
**Voltage Gain**

$$\frac{U_2}{U_1} = -\frac{h_{21}}{\det h}$$

Where the quantities are as defined in *Table 26*.

**Formula 100.4**
**Penetration Coefficient**

$$\frac{U_1}{U_2} = -\frac{\det h}{h_{21}}$$

Where the quantities are as defined in *Table 26*.

**Formula 100.5**
**Direct-Transference Resistance**

$$\frac{U_2}{U_1} = -\frac{h_{21}}{h_{22}}$$

Where the quantities are as defined in *Table 26*.

# 101. SHORT-CIRCUIT INPUT

$R_g = 0$ and $U_1 = 0$

The next formulas are used to determine the characteristic quantities in the three basic configurations.

**Formula 101.1**
**Output Resistance**

$$\frac{U_2}{I_2} = \frac{h_{11}}{\det h}$$

Where the quantities are as defined in *Table 26*.

## Formula 101.2
## Output Conductance

$$\frac{I_2}{U_2} = \frac{\det h}{h_{11}}$$

Where the quantities are as defined in *Table 26*.

## Formula 101.3
## Feedback Resistance

$$\frac{U_2}{I_1} = -\frac{h_{11}}{h_{12}}$$

Where the quantities are as defined in *Table 26*.

## Formula 101.4
## Fedback Conductance

$$\frac{I_1}{U_2} = -\frac{h_{12}}{h_{11}}$$

Where the quantities are as defined in *Table 26*.

# 102. OPEN-CIRCUIT INPUT

$$R_g = \infty \qquad I_1 = 0$$

(Rg = generator resistance)

The following formulas are used to determine the characteristics of the three transistor configurations.

**Formula 102.1**

**Output Resistance**

$$\frac{U_2}{I_2} = \frac{1}{h_{22}}$$

Where the quantities are as defined in *Table 26*.

**Formula 102.2**

**Output Conductance**

$$\frac{I_2}{U_2} = h_{22}$$

Where the quantities are as defined in *Table 26*.

**Formula 102.3**

**Voltage Feedback**

$$\frac{U_1}{U_2} = h_{12}$$

Where the quantities are as defined in *Table 26*.

**Formula 102.4**

**Transfer Inverse Resistance**

$$\frac{U_1}{I_2} = \frac{\cdot h_{12}}{h_{22}}$$

Where the quantities are as defined in *Table 26*.

# 103. COMMON-BASE CONFIGURATION USUAL FORMULAS

*Figure 88* shows the basic configuration where Rg is the internal resistance generator and $R_L$ the load resistance.

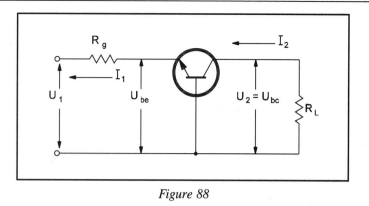

*Figure 88*

NOTE: The index "b" used with the hybrid parameters to indicate they are common-base was supressed in the next formulas for easier use. Instead of $h_{21e}$ we are using $h_{21}$.

## Formula 103.1
## Current Gain

$$Gi = \frac{I_c}{I_e}$$

$$G_i = \frac{h_{21}}{1 + h_{22}xR_L}$$

Where:

The quantities are as defined in *Table 26*

RL is the load resistance in ohms ($\Omega$)

## Formula 103.2
## Voltage Gain

$$G_v = \frac{U_{bc}}{U_{be}}$$

$$G_v = \frac{h_{21}xR_L}{h_{11} + \det hxR_L}$$

Where:

$G_V$ is the voltage gain

Other quantities are as defined in *Table 26*

$R_L$ is the load resistance in ohms ($\Omega$)

## Formula 103.3
## Input Resistance

$$R_{in} = \frac{U_{bc}}{I_C}$$

$$R_{in} = \frac{h_{11} + \det hxR_L}{1 + h_{22}xR_L}$$

Where:

Rin is the input resistance in ohms ($\Omega$)

Other quantities are as defined as in *Table 26*

## Formula 103.4
## Output Resistance

$$R_{OUT} = \frac{U_{bc}}{I_e}$$

$$R_{OUT} = \frac{h_{11} + R_G}{\det h + h_{22}xR_G}$$

Where:

Rout is the output resistance in ohms ($\Omega$)

$R_G$ is the generator resistance

The other quantities are as defined in *Table 26*

## Formula 103.5
## Power Gain

$$G_P = GixG_U$$

$$G_P = \frac{h_{21}^2 x R_L}{\left(1 + h_{21} x R_L\right) x \left(h_{11} + \det hxR_L\right)}$$

Where:

Gp is the power gain

Gi is the current gain

Gv is the voltage gain

$R_L$ is the load resistance in ohms ($\Omega$)

Other parameters are as defined by *Table 26*

## Derivated Formulas:

## Formula 103.6

When the input impedance is matched with the input resistance (Ri = $R_G$) the following formula is used to calculate the power gain:

$$G_P = \frac{4xR_G x R_L x h_{21}^2}{\left[R_G\left(1 + h_{22} x R_L\right) + \left(h_{11} + \det hxR_L\right)\right]^2}$$

Where:

Gp is the power gain

Rg is the generator resistance in ohms ($\Omega$)

$R_L$ is the load resistance in ohms ($\Omega$)

Other parameters are as defined by *Table 26*

# 104. COMMON-EMITTER CONFIGURATION USUAL FORMULAS

When used in the common-emitter configuration a transistor is wired as shown in *Figure 89*. RL is the load resistance and Rg is the generator resistance.

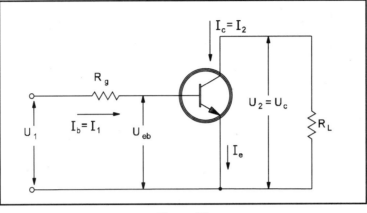

*Figure 89*

NOTE: The index "e" used with the hybrid parameters to indicate they are common-emitter was supressed in the next formulas for easier use. So instead of $h_{21e}$ we are using $h_{21}$.

## Formula 104.1

### Current Amplification

$$G_I = \frac{I_C}{I_b}$$

$$G_I = \frac{h_{21}}{1 + h_{22} x R_L}$$

Where:

G$_I$ is the current gain

The other parameters are as defined in *Table 26*

## Formula 104.2
## Voltage Gain

$$G_V = \frac{U_{ec}}{U_{eb}}$$

$$G_V = \frac{h_{21}xR_L}{h_{11} + \det hxR_L}$$

Where:

Gv is the voltage gain

Other parameters are as defined in *Table 26*

## Formula 104.3
## Input Resistance

$$R_{IN} = \frac{U_{eb}}{I_e}$$

$$R_{IN} = -\frac{h_{11} + \det hxR_L}{1 + h_{22}xR_L}$$

Where:

Rin is the input resistance in ohms ( $\Omega$ )

Other parameters are defined in *Table 26*

## Formula 104.4
## Output Resistance

$$R_{OUT} = \frac{U_{ec}}{I_c}$$

$$R_{OUT} = \frac{h_{11} + R_G}{\det h + h_{22}xR_G}$$

Where:

Rout is the output resistance in ohms ($\Omega$)

Other parameters are as defined as in *Table 26*

## Formula 104.5
## Power Gain

$$Gp = \frac{h_{11}^2 x R_L}{(1 + h22 + R_L) x (h_{11} + \det hxR_L)}$$

Where:

Gp is the power gain

The other quantities are as defined as in *Table 26*

## Formula 104.6
## Power Gain with Impedance Matching

$$G_P = \frac{4 x R_G x R_L x h_{21}^2}{\left[ R_G (1 + h_{22} x R_L) + (h_{11} + \det hxR_L) \right]^2}$$

Where:

Gp is the power gain

Other parameters are as defined in *Table 26*

# 105. COMMON-COLLECTOR USUAL FORMULAS

The circuit used in this configuration is shown in *Figure 90*. $R_G$ is the generator resistance and $R_L$ is the load resistance.

NOTE: The index "c" used with the hybrid parameters to indicate they are common-collector was supressed in the next formulas for ease of use. So instead of $h_{22c}$ we are using $h_{22}$.

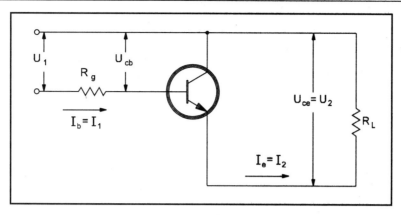

*Figure 90*

# Formula 105.1
# Current Gain

$$G_I = \frac{I_e}{I_b}$$

$$G_I = \frac{h_{22}}{1 + h_{22} x R_L}$$

Where:

G$_I$ is the current gain

Other parameters are as defined in *Table 26*

# Formula 105.2
# Voltage Gain

$$G_V = \frac{U_{ce}}{U_{cn}}$$

$$G_V = \frac{h_{21} x R_L}{h_{11} + \det h x R_L}$$

Where:

Gv is the voltage gain

Other parameters are as defined in *Table 26*

## Formula 105.3
## Input Resistance

$$R_{IN} = \frac{U_{cb}}{I_e}$$

$$R_{IN} = \frac{h_{11} + \det h x R_L}{1 + h_{22} x R_L}$$

Where:

$R_{IN}$ is the input resistance in ohms ($\Omega$)

Other parameters are as defined in *Table 26*

## Formula 105.4
## Output Resistance

$$R_{OUT} = \frac{U_{ce}}{I_e}$$

$$R_{OUT} = \frac{h_{11} + R_G}{\det h + h_{22} x R_G}$$

Where:

$R_{OUT}$ is the output resistance in ohms ($\Omega$)

Other parameters are as defined in *Table 26*

## Formula 105.5
## Power Gain

$$G_P = \frac{h_{21}^2 x R_L}{\left(1 + h_{22} x R_L\right) x \left(h_{11} + \det h x R_L\right)}$$

Where:

$G_P$ is the power gain

Other parameters are as defined in *Table 26*

## Formula 105.6

## Power Gain with Impedance Matching

$$G_P = \frac{4 x R_G x R_L x h_{21}^2}{\left[ R_G \left( 1 + h_{22} x R_L \right) x \left( h_{11} + \det h x R_L \right) \right]^2}$$

Where:

$G_P$ is the power gain

Other parameters are as defined in *Table 26*

## TABLE 28

## Other Names of Symbols used in Transistor Specifications

| Hybrid Parameter Name | Term | Other Symbol |
|---|---|---|
| $h_{11}$ | Input impedance | $h_i$ |
| $h_{12}$ | Reverse voltage ratio | $h_r$ |
| $h_{21}$ | Forward current ratio | $h_e$ |
| $h_{22}$ | Output conductance | $h_o$ |

## TABLE 29

## Other Symbols for Common-Emitter Configuration

| Hybrid Parameter Name | Term | Symbol | Other Symbol |
|---|---|---|---|
| $h_{11}$ | Input impedance | $h_{ie}$ | $r_a$ |
| $h_{12}$ | Reverse voltage ratio | $h_{re}$ | - |
| $h_{21}$ | Forward current ratio | $h_{fe}$ | $\beta$ |
| $h_{22}$ | Output conductance | $h_{oe}$ | $\dfrac{1}{r_s}$ |

**TABLE 30**

**Other Symbols for Common-Base Configuration**

| Hybrid Parameter Name | Term | Symbol | Other Symbol |
|---|---|---|---|
| $h_{11}$ | Input impedance | $h_{ib}$ | $r_b$ |
| $h_{12}$ | Reverse voltage ratio | $h_{rb}$ | - |
| $h_{21}$ | Forward current ratio | $h_{fb}$ | $\alpha$ |
| $h_{22}$ | Output conductance | $h_{ob}$ | $\dfrac{1}{r_s}$ |

**TABLE 31**

**Other Symbols for Common-Collector Configuration**

| Hybrid Parameter Name | Term | Symbol | Other Symbol |
|---|---|---|---|
| $h_{11}$ | Input impedance | $h_{ic}$ | $r_c$ |
| $h_{12}$ | Reverse voltage ratio | $h_{rc}$ | - |
| $h_{21}$ | Forward current ratio | $h_{fc}$ | $\delta$ |
| $h_{22}$ | Output conductance | $h_{oc}$ | $\dfrac{1}{r_s}$ |

# TRANSISTOR PRACTICAL FORMULAS

Simplified formulas are useful in many practical projects where bipolar transistors are the core element. Next are practical formulas that can be used in noncritical projects where the use of hybrid parameters are not obligatory.

# 106. LOAD RESISTANCE

The load resistance of a transistor depends on the power supply voltage, the voltage between collector and emitter, and also the amount of current flowing to the collector. *Figure 91* shows the typical family of characteristic curves of a transistor and the basic circuit to calculate the load resistance using the following formula.

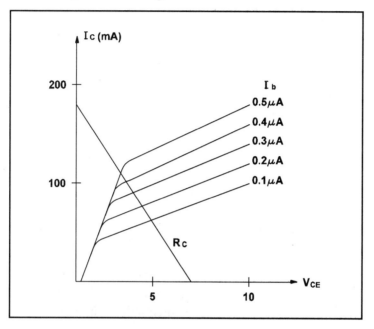

*Figure 91*

## Formula 106.1
## Load Resistance

$$R_C = \frac{U_b - U_{ce}}{I_c}$$

Where:

Rc is the load resistance in ohms ($\Omega$)

$U_b$ is the power supply voltage in volts (V)

$U_{ce}$ is the voltage between collector and emitter in volts (V)

Ic is the collector current in amperes (A)

**Application Example:**

In a circuit a transistor must operate within the following conditions: voltage between collector and emitter = 3.0V; collector current = 50 mA; power supply voltage = 12V.

Data:

Uce = 3.0V

Ic = 50 mA = 0.05 A

Ub = 12V

Applying Formula 106.1:

$$Rc = \frac{12 - 3}{0.05} = \frac{9}{0.05} = 180 \text{ ohms}$$

# 107. BASE-BIASING RESISTANCE

When calculating the base resistance of a transistor in the common-emitter configuration it is important to keep the collector voltage near the center of the load line, as shown in *Figure 92* to avoid distortion of signals. The biasing resistance will depend on the power supply voltage and the base current.

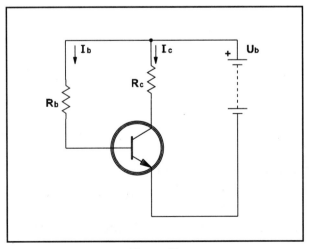

*Figure 92*

## Formula 107.1
## Base Resistor

$$R_b = \frac{U_b}{I_b}$$

Where:

    $R_b$ is the base resistance on ohms ($\Omega$)

    Ub is the power supply voltage in volts (V)

    Ib is the base current in amperes (A)

**Application Example:**

Find the base resistance of a transistor with a load resistance of 1 kilohm and a collector current of 10 mA. A 12V power supply is used in this circuit.

*Figure 93* shows the characteristic curves for the transistor used.

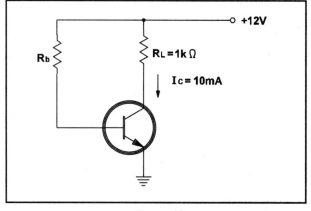

*Figure 93*

Data:

    $R_L$ = 1000 ohms

    Ic = 10 mA = 0.01 A

    Ub = 12V

From the characteristc curves we can see that when the collector current is 10 mA the current flowing to the base is 50 uA.

Applying Formula 107.1:

$$Rb = \frac{12}{50x10^{-6}} = \frac{12}{50}x10^6 = 0.24x1$$

# 108. AUTOMATIC-BIASING RESISTANCE

To increase the performance of a common-emitter amplifier using a transistor, the base-biasing resistor can be wired as shown in *Figure 94*. The following formula is used to calculate this resistor.

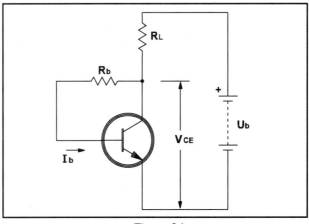

*Figure 94*

**Formula 108.1**
**Base Resistor**

$$R_B = \frac{U_{CE}}{I_B}$$

Where:

$R_B$ is the base resistance in ohms ($\Omega$)

$U_{CE}$ is the voltage between collector and emitter in volts (V)

$I_B$ is the base current in amperes (A)

## Application Example:

Determine the base-biasing resistor in a circuit where the collector voltage is 6V and base current is 50 $\mu$ A.

Data:

$$Uce = 6V$$

$$I_B = 50 \ \mu A$$

Applying Formula 108.1

$$R_B = \frac{6}{50x10^{-6}} = \frac{6}{50}x10^6 = 0.12x10^6 = 120 k\Omega$$

# JUNCTION FIELD-EFFECT TRANSISTOR (JFET) AND MOS FIELD-EFFECT TRANSISTOR (MOSFET)

The Junction Field-Effect Transistor and the Metal-Oxide Field-Effect Transistor can be used in three basic configurations. The following formulas describe the performance of FETs and MOSFETs in these configurations.

# 109. COMMON SOURCE

In the common-source configuration the signal is applied to the gate and taken from the drain as shown in *Figure 95*.

## Formulas 109.1
## Common-Source Gain

$$Gv = \frac{U_{OUT}}{U_{in}}$$

$$Gv = -gmxR_L$$

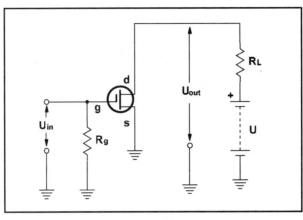

*Figure 95*

Where:

Gv is the voltage gain

Uout is the output voltage in volts (V)

Uin is the input voltage in volts

gm is the transconductance in Siemens (S)

$R_L$ is the load resistance in ohms ($\Omega$)

**Application Example:**

Determine the voltage gain in a circuit using a JFET with 1000 $\mu$ S of transconductance and a 10 kilohm load resistance.

Data:

$$gm = 1\,000\ \mu S = 10^{-3}\ S$$

$$R_L = 10\ k = 10 \times 10^3\ ohms$$

Applying Formula 109.1:

$$Gv = -10^{-3}x10x10^3 = -10$$

# 110. COMMON-DRAIN

This configuration is shown in *Figure 96*. The signal is applied to the gate and taken from the source.

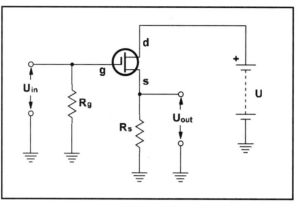

*Figure 96*

## Formulas 110.1
## Common-Drain Gain

$$Gv = \frac{U_{OUT}}{U_{IN}}$$

$$Gv = \frac{gmxRs}{1 + gmxRs}$$

Where:

Gv is the voltage gain

Uout is the output voltage in volts (V)

Uin is the input voltage in volts (V)

gm is the transconductance in siemens (S)

Rs is the source resistance in ohms ($\Omega$)

NOTE: The value for of Gv in common applications is generally near 1.

## Formula 110.2
## Output Impedance

$$R_{OUT} = \frac{R_S}{1 + gmxR_S}$$

Where:

Rout is the output impedance in ohms ($\Omega$)

Rs is the source resistance in ohms ($\Omega$)

gm is the transconductance in siemens (S)

**Application Example:**

Calculate the voltage gain and the output impedance for a common-source circuit using a JFET with 5000 uS of transconductance and a 5000-ohm source resistor.

Data:

Rs = 5 000 ohms = 5 x $10^3$ ohms

gm = 5 000 $\mu$ S = 5 x $10^{-6}$ S

Gv = ?

Rout = ?

Calculating Gv, using Formula 110.1:

$$Gv = \frac{5x10^{-3}x5x10^3}{1 + 5x10^{-3}x5x10^3} = \frac{25}{1 + 25} =$$

Calculating Rout, using Formula 110.2:

$$Rout = \frac{5x10^3}{1 + 5x10^{-3}x5x10^3} = \frac{5000}{6} = 833.3 \text{ ohms}$$

# 111. COMMON-GATE

In the common-gate configuration, a FET is wired as shown in *Figure 97*. The signal is applied to the source and taken from the drain.

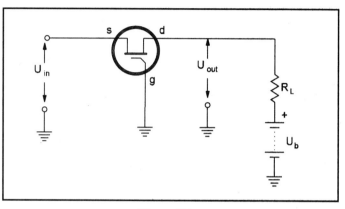

*Figure 97*

## Formula 111.1
## Common-Gate Gain

$$Gv = gmxR_L$$

Where:

Gv is the voltage gain

gm is the transconductance in Siemens (S)

$R_L$ is the load resisance in ohms ($\Omega$)

## Formula 111.2

$$R_{IN} = \frac{1}{gm}$$

Where:

Rin is the input resistance in ohms ($\Omega$)

gm is the transconductance in Siemens (S)

## Application Example:

What is the voltage gain and input impedance of a circuit using a JFET in common-gate configuration where the load resistance is 10 kohms? The transconductance of the transistor used is 1000 $\mu$ S.

Data:

$$R_L = 10 \, k\Omega$$

$$gm = 1 \times 10^{-3} \, S$$

$$Gv = ?$$

$$Rin = ?$$

Calculating Gv, using Formula 111.1:

$$Gv = 1x10^{-3}x10x10^{3} = 10$$

Determining Rin, using Formula 111.2:

$$Rin = \frac{1}{10^{-3}} = 10^{3} = 1000 \text{ ohms}$$

# 112. UNIJUNCTION TRANSISTOR (UJT)

The unijunction transistor (UJT) and also the programmable UJT are devices intended to be used in relaxation oscillators and timers. Figure 98 shows the equivalent circuit and the basic configuration of a relaxation oscillator using the UJT.

## Formula 112.1
## UJT Peak Voltage

$$U_P = U_D + \eta x U_{BB}$$

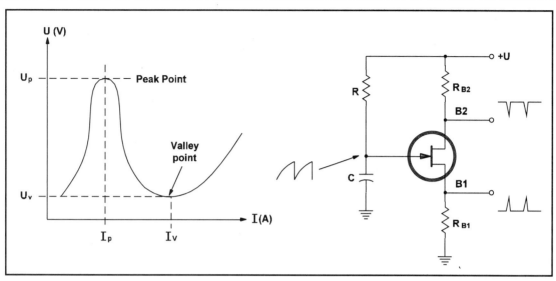

*Figure 98*

Where:

Up is the peak voltage in volts (V)

$U_D$ is the foward-biased voltage fall in the UJT internal diode (O.7V)

$\eta$ is the intrinsic standoff ratio (0.3 to 0.8 in typical UJT transistors)

$U_{BB}$ is the voltage between bases in volts (V)

## Formula 112.2
## Resistance Between Bases

$$R_{BB} = R_{B1} + R_{B2}$$

Where:

$R_{BB}$ is the resistance between bases in ohms ($\Omega$)

$R_{B1}$ and $R_{B2}$ are the internal equivalent resistances in ohms ($\Omega$)

# Formula 112.3
## As Relaxation Oscillator

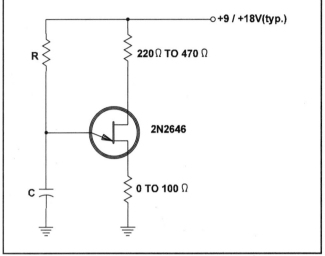

*Figure 99*

Frequency:

$$f = \frac{1}{RxCx \ln\left(\dfrac{1}{1-\eta}\right)}$$

Where:

f is the frequency in hertz (Hz)

R is the resistance in ohms ($\Omega$)

C is the capacitance in farads (F)

$\eta$ is the intrinsic standoff ratio

## Derivated Formula:

## Formula 112.4

A simplified formula to calculate the frequency in a relaxation oscillator can be used in applications where precision is not required.

$$f = \frac{1}{0.82\,xRxC}$$

Where:

    f is the frequency in hertz (Hz)

    R is the resistance in ohms ($\Omega$)

    C is the capacitance in farads (F)

## Formula 112.5
## When the Circuit is Used as Timer

$$T = 0.82\,xRxC$$

Where:

    T is the period in seconds (s)

    R is the resistance in ohms ($\Omega$)

    C is the capacitance in farads (F)

**Application Example:**

Determine the resistance to be used in a relaxation oscillator using a UJT to make it run at 1 kHz with a 100 nF capacitor.

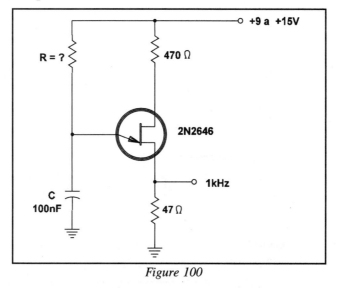

*Figure 100*

Data:

$$f = 1\ 000\ Hz\ 10^3$$

$$C = 100\ nF = 0.1 \times 10^{-6}\ F$$

$$R = ?$$

Using Formula 112.4:

$$10^3 = \frac{1}{0.82 x 0.1 x 10^{-6} x R}$$

Isolating R:

$$R = \frac{1}{0.82 x 0.1 x 10^{-6} x 10^3}$$

$$R = \frac{1}{0.082 x 10^{-3}} = \frac{1 x 10^3}{0.082} = 12.19 x 10^3 = 12.19\ k\Omega$$

# 113. SCR

SCRs or Silicon Controlled Rectifiers are four-layer semiconductor devices intended to be used as power controls and relaxation oscillators. The symbol and characteristic of this component are shown in *Figure 101*. Since the SCR is triggered by external devices in a on/off mode, there are few quantities to be calculated when using them in practical projects.

## Formula 113.1
## Trigger Current

$$I_T = \frac{I_{co}}{1 - \alpha 1 + \alpha 2}$$

Where:

$I_T$ is the trigger current in amperes (A)

Ico is the leakage current in amperes (A)

$\alpha 1$ is the gain of first transistor

$\alpha 2$ is the gain of second transistor

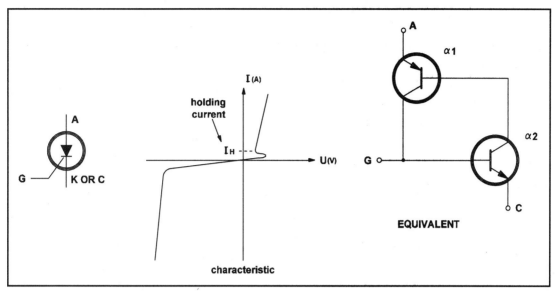

*Figure 101*

NOTE: The parameter $I_T$ usually is given by the SCR's manufacturer and is in the range between 0.1 and 100 mA for common types.

## Formula 113.2
## Power per Cycle

$$Pd = \frac{U_A x I_A x tr}{4.6}$$

Where:

    Pd is the power dissipation per cycle in watts (W)

    $U_A$ is the anode voltage prior to switching in volts (V)

    $I_A$ is the anode current after switching in amperes (A)

    tr is the switching time for anode-cathode voltage to fall from 90% to 10%

## Formula 113.3
## Average Power

$$Pd = \frac{f x U_A x I_A x tr}{4.6}$$

Where:

Pd is the average power dissipation in watts (W)

f is the switching frequency in hertz (Hz)

$U_A$ is the anode voltage prior to switching in volts (V)

$I_A$ is the anode current after switching in amperes (A)

## Formula 113.4
## DC Application

$$Pd = Us x Id$$

Where:

Pd is the dissipated power in watts (W)

Us is the voltage fall with the SCR in the on-state, also called saturation voltage, in volts (V)

Id is the forward current in amperes (A)

NOTE: For common SCRs the typical value for Uf is 2.0V.

## Formula 113.5
## Load voltage vs. Trigger Delay (angle $\alpha$ ), Half-Wave Application

$$UL = \frac{Up}{2 x \pi} x (1 + \cos\alpha)$$

Where:

UL is the voltage in the load (instantaneous) in volts (V)

Up is the peak voltage of the sine input voltage in volts (V)

$\pi$ is the constant 3.1416

cos $\alpha$ is the cosine of the conduction angle in degrees

## Derivated Formulas:

## Formula 113.6
## Load Voltage for Full-Wave Application

$$UL = \frac{Up}{\pi} x(1 + \cos\alpha)$$

Where:

$U_L$ is the instantaneous value of the load voltage in volts (V)

Up is the peak value of the sine wave input voltage in volts (V)

$\pi$ is 3.1416

$\cos\alpha$ is the cosine of the conduction angle in degrees

**Application Example:**

An SCR is used to control a DC load in a 5A circuit. What is the power dissipated by this device? Saturation voltage = 2.0V.

Data:

Us = 2.0V

Id = 5A

Pd = ?

Using Formula 113.4:

Pd = 2.0 x 5 = 10 W

# 114. TRIAC

The triac is a semiconductor switching device that can be considered as two SCRs connected in a manner that allows them to pass alternating current, as shown in *Figure 102*. So this device can be treated like two SCRs when making calculations.

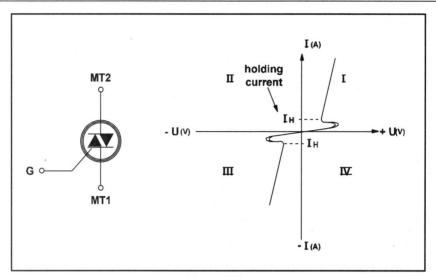

*Figure 102*

## Formula 114.1
## Power Dissipation

$$Pd = U_S x I_d$$

Where:

Pd is the power dissipation in watts (W)

Us is the saturation voltage in volts (V)

Id is the forward current in amperes (A)

NOTE: For common triacs the saturation voltage is around 1.5 volts.

## Formula 114.2

Formula 113.6, used for SCRs when applied to full-wave circuits, is valid for calculations involving the load voltage as function of the conduction angle.

$$UL = \frac{Up}{\pi} x (1 + \cos\alpha)$$

Where:

UL is the instantaneous value of the load voltage in volts (V)

Up is the peak value of the sine input voltage in volts (V)

$\pi$ is 3.1416

$\cos \alpha$ is the cosine of the conduction angle (angle in degrees)

**Application Example:**

Calculate the power dissipated by a triac when used in a circuit controlling a 10A load. The saturation voltage is 1.5V.

Data:

Us = 1.5 V

Id = 10 A

Pd = ?

Using Formula 114.1:

Pd = 10 x 1.5 = 15 watts

# OSCILLATORS

# 115. ASTABLE MULTIVIBRATOR

The astable multivibrator is formed by two bipolar transistors wired as shown in *Figure 103*. This circuit produces a square wave with duty cycle depending on the conduction time of each transistor. Equivalent configurations can be made based on FETs and even tubes.

**Formula 115.1**
**Conduction Time**

$$tp = 0.69xRxC$$

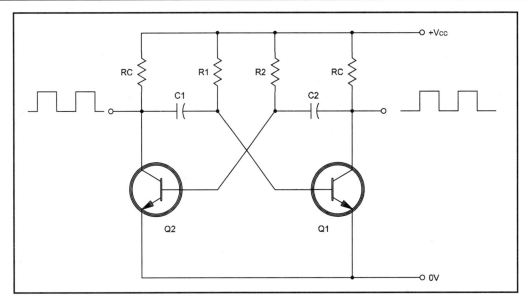

*Figure 103*

Where:

tp is the conduction time of one transistor in seconds (s)

R is the resistance in ohms ($\Omega$ )

C is the capacitance in farads (F)

NOTE: For Q1 the conduction time tp1 is determined by R1 and C1, and for Q2 the conduction time tp2 by R2 and C2.

## Formula 115.2
## Frequency

$$f = \frac{1}{tp1 + tp2}$$

$$f = \frac{1}{0.69x(R1xC1 + R2xC2)}$$

Where:

f is the frequency in hertz (Hz)

R1, R2 are the resistances in ohms ($\Omega$)

C1, C2 are the capacitances in farads (F)

## Derivated Formulas:

### Formula 115.3
### Square Wave Oscillator (50% Duty Cycle)

R1 = R2 = R and C1 = C2 = C

$$f = \frac{1}{1.38xRxC}$$

Where:

      f is the frequency in hertz  (Hz)

      R is the resistance in ohms ($\Omega$)

      C is the capacitance in farads (F)

### Formula 115.4
### Frequency as Function of Conduction Time

R1 = R2 = R and C1 = C2 = C

$$f = \frac{1}{2xtp}$$

Where:

      f is the frequency in hertz (Hz)

      tp is the conduction time in seconds (s)

**Application Example:**

Determine the frequency of the free-runing multivibrator with the schematic diagram shown in *Figure 104*.

Data:

      R1 = R2 = 100 k$\Omega$ = 10 x $10^3$ $\Omega$

      C1 = C2 = 10 nF = 0.01 $\mu$ F x $10^{-6}$

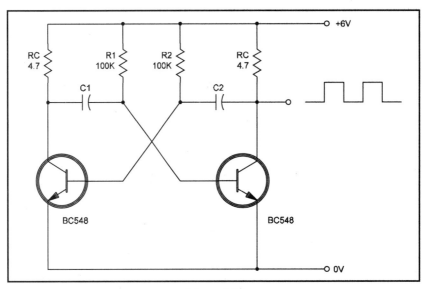

*Figure 104*

Using formula 115.3:

$$f = \frac{1}{1.38x10x10^3x0.01x10^{-6}} \qquad = 7\ 246\ \text{kHz}$$

# 116. NEON-LAMP OSCILLATOR

*Figure 105* shows the basic configuration of a relaxation oscillator using a neon lamp. This circuit is suitable for frequencies up to some tens of kilohertz and operates from power supply voltages over 80 volts.

A typical neon lamp breaks down at 70V and establishes a glow, and this glow is extinguished when the voltage across the lamp is reduced to 50V. The next formula is valid for these parameters.

## Formula 116.1
## Period

$$T = RxCx\ln\left(\frac{U - Uh}{U - Ut}\right)$$

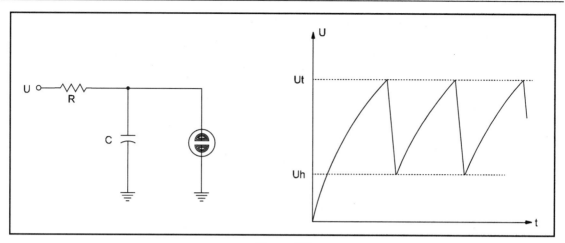

*Figure 105*

Where:

T is the period in seconds (s)

C is the capacitance in farads (F)

R is the resistance in ohms ($\Omega$)

U is the power supply voltage in volts (V)

Ut is the triggering voltage in volts (V) (70V typ.)

Uh is the holding voltage in volts (V) (60V typ.)

## Formula 116.2
## Frequency

$$f = \frac{1}{RxCx\ln\left(\dfrac{U-Uh}{U-Ut}\right)}$$

Where:

f is the frequency in hertz (Hz)

C is the capacitance in farads (F)

U is the power supply voltage in volts (V)

Ut is the triggering voltage in volts (V)

Uh is the holding voltage in volts (V)

## Application Example:

In the circuit shown in *Figure 106* the resistance between gate and cathode of the SCR in the ON state is not considered. Determine the frequency of the produced flashes:

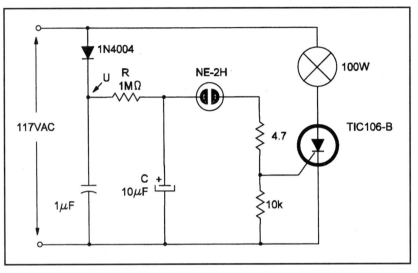

*Figure 106*

Data:

$$R = 1\,M\Omega = 10^6\,\Omega$$

$$C = 10\,\mu F = 10 \times 10^{-6}\,F$$

$$U = 150V$$

$$Ut = 70V$$

$$Uh = 50V$$

$$f = ?$$

Applying Formula 116.2:

$$f = \cfrac{1}{10^6 x10x10^{-6}x\ln\left(\cfrac{150-50}{150-70}\right)} = \cfrac{1}{10x\ln\left(\cfrac{100}{80}\right)} = \cfrac{1}{10x\ln(1.25)} = \cfrac{1}{10x0.223} = \cfrac{1}{2.23} = 0.448\ f$$

$$= 0.448\ Hz \quad \text{(about two flashes per second)}$$

# 117. PHASE-SHIFT OSCILLATOR

This circuit produces low-frequency sine wave signals. The range is up to some hundreds of kilohertz and any NPN or PNP general-purpose transistor can be used in the basic configuration. The following formula is used to calculate the frequency:

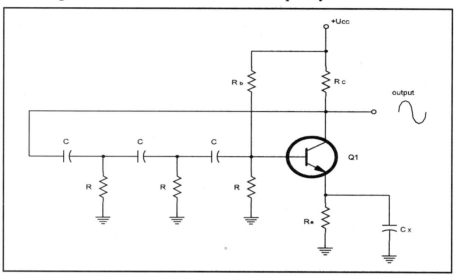

*Figure 107*

## Formula 117.1
## Frequency

$$f = \frac{1}{4.88 \times \pi \times R \times C}$$

Where:

f is the frequency in hertz (Hz)

$\pi$ is the constant 3.1416

R is the resistance in ohms ($\Omega$)

C is the capacitance in farads (F)

NOTE: 4.88 = 2 x $\sqrt{6}$

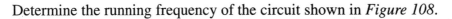

## Application Example:

Determine the running frequency of the circuit shown in *Figure 108*.

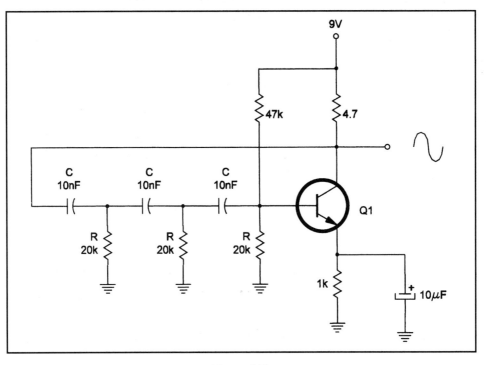

*Figure 108*

Data:

$C = 10$ nF $= 0.01$ $\mu$F

$R = 20$ k$\Omega$

$f = ?$

Applying formula 117.1:

$$f = \frac{1}{4.88 x 3.14 x 20 x 10^3 x 0.01 x 10^{-6}} = \frac{10^3}{3.06} = 326.8 \text{ Hz}$$

# 118. WIEN BRIDGE OSCILLATOR

The basic configuration using a transistor is shown in *Figure 109*. This circuit produces a sine wave low-frequency signal in the frequency range between some hertz and some hundreds of kilohertz. In practical applications it is recommended to make C1 = C2 and R1 = R2.

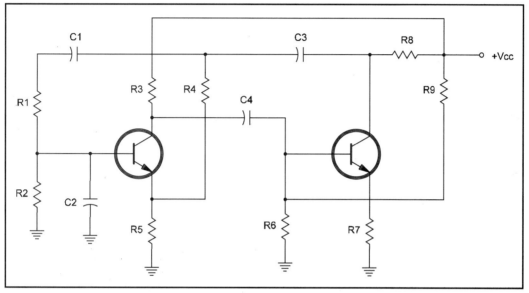

*Figure 109*

## Formula 118.1
## Frequency

$$f = \frac{1}{2 \times \pi \times \sqrt{R1 \times R2 \times C1 \times C2}}$$

Where:

f is the frequency in hertz (Hz)

R1 and R2 are the resistances in ohms ( $\Omega$ )

C1 and C2 are the capacitances in farads (F)

$\pi$ is the constant 3.1416

**Application Example:**

Calculate frequency of a wien-bridge oscillator: C1 = C2 = 20 nF and R1 = R2 = 20 kilohms.

Data:

$$R1 = R2 = 20 \text{ k}\Omega$$

$$C1 = C2 = 20 \text{ nF}$$

Using Formula 118.1:

$$f = \frac{1}{2x3.14x\sqrt{20x10^3\,x20x10^3\,x20x10^{-9}\,x20x10^{-9}}}$$

$$f = \frac{1}{6.28x\sqrt{160x10^3x10^{-12}}}$$

$$f = \frac{1}{6.28x\sqrt{160x10^{-9}}} = \frac{1}{6.28x\sqrt{16x10^{-8}}} = \frac{1}{6.28x4x10^{-4}} = \frac{10^4}{6.28}$$

$$f = 1\ 592.35 \text{ Hz or } 1.592 \text{ kHz}$$

# 119. TWIN-T OSCILLATOR

The twin-T oscillator is used to produce low-frequency sine wave signals and damped oscillations in the audio range. Damped oscillations are specially used to generate percussion sounds in electronic musical instruments.

*Figure 110* shows the basic configuration with common values of resistors in a twin-tee oscillator using an NPN general-purpose transistor.

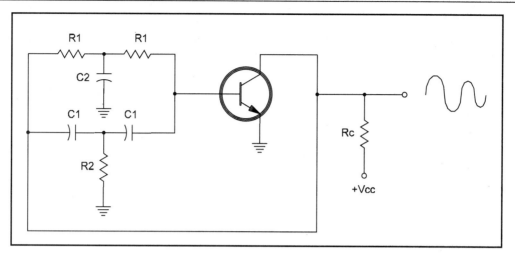

*Figure 110*

## Formulas 110.1
## Frequency

$$f = \frac{1}{2 \times \pi \times R_1 \times C_2}$$

$$f = \frac{1}{2 \times \pi \times R_2 \times C_1}$$

Where:

f is the frequency in hertz (Hz)

$\pi$ is the constant 3.1416

R1, R2 are resistances in ohms ($\Omega$)

C1, C2 are the capacitances in farads (F)

**IMPORTANT:**

C1, C2 and R1 and R2 must mantain the following ratios of values:

**C1 = C2/2**

**R1 = 2 x R2**

Damping oscillations can be generated by reducing R2. A simple away to produce the desired damping oscillation or fixing the damping time is by replacing R2 with an adjustable resistor (trimmer potentiometer, for instance). The value is the same found in the calculations.

**Application Example:**

Calculate the oscillation frequency of the oscillator shown in *Figure 110* when C1 = 100 nF and C2 =200 nF.

Data:

R2 = 100 k$\Omega$

R1= 50 k$\Omega$

C1 = 100 nF

C2 = 200 nF

f = ?

Using Formula 119.1:

$$f = \frac{1}{2x3.14x50x10^3x200x10^{-9}} = \frac{1}{6.28x10^6x10^{-9}} = \frac{1x10^3}{6.28} = 159.23 \text{ Hz}$$

## TABLE 32
## Greek Alphabet

| Name | Large | Small | English Equivalent |
|------|-------|-------|--------------------|
| Alpha | A | $\alpha$ | a |
| Beta | B | $\beta$ | b |
| Gamma | $\Gamma$ | $\gamma$ | g |
| Delta | $\Delta$ | $\delta$ | d |
| Epsilon | E | $\varepsilon$ | c (short as in "met") |
| | (continued next page) | | |

| Name | Large | Small | English Equivalent |
| --- | --- | --- | --- |
| Zeta | Z | $\zeta$ | z |
| Eta | H | $\eta$ | e (long as in "meet") |
| Theta | $\Theta$ | $\theta$ | th |
| Iota | I | $\iota$ | i |
| Kappa | K | $\kappa$ | k |
| Lambda | $\Lambda$ | $\lambda$ | l |
| Mu | M | $\mu$ | m |
| Nu | N | $\nu$ | n |
| Xi | $\Xi$ | $\xi$ | x |
| Omicron | O | $o$ | o (as in "olive") |
| Pi | $\Pi$ | $\pi$ | p |
| Rho | P | $\rho$ | r |
| Sigma | $\Sigma$ | $\sigma$ | s |
| Tau | T | $\tau$ | t |
| Upsilon | Y | $\upsilon$ | u |
| Phi | $\Phi$ | $\varphi$ | ph |
| Chi | X | $\chi$ | ch (as in "school") |
| Psi | $\Psi$ | $\phi$ | ps |
| Omega | $\Omega$ | $\omega$ | o (as in "hole") |

## TABLE 33

### Equally Tempered Chromatic Scale Frequencies (Hz)

This scale can be used as reference when projecting percussion sound generators based in the twin-tee oscillator.

| | | | | | | | | | |
|---|---|---|---|---|---|---|---|---|---|
| **B** | 30.867 | 61.735 | 123.47 | 246.94 | 493.88 | 987.77 | 1975.53 | 3951.07 | 7902.13 |
| **A#** | 29.135 | 58.270 | 116.54 | 233.08 | 466.16 | 932.32 | 1864.66 | 3729.31 | 7458.63 |
| **A** | 27.500 | 55.000 | 110.00 | 220.00 | 440.00 | 880.00 | 1760.00 | 3520.00 | 7040.00 |
| **G#** | 25.956 | 51.913 | 103.82 | 207.65 | 415.31 | 830.61 | 1661.22 | 3322.44 | 6644.88 |
| **G** | 24.449 | 48.999 | 97.988 | 195.99 | 391.99 | 783.99 | 1567.98 | 3135.97 | 6271.93 |
| **F#** | 23.124 | 46.249 | 92.499 | 184.99 | 369.99 | 739.99 | 1479.98 | 2959.96 | 5919.92 |
| **F** | 21.826 | 43.653 | 87.307 | 174.61 | 349.23 | 698.46 | 1396.91 | 2793.83 | 5587.66 |
| **E** | 20.601 | 41.203 | 82.406 | 164.81 | 329.63 | 659.26 | 1318.51 | 2637.02 | 5274.04 |
| **D#** | 19.445 | 38.890 | 77.781 | 155.56 | 311.13 | 622.25 | 1244.51 | 2489.02 | 4978.03 |
| **D** | 18.354 | 36.708 | 73.416 | 146.83 | 293.66 | 587.33 | 1174.66 | 2349.32 | 4698.64 |
| **C#** | 17.324 | 34.648 | 69.295 | 138.59 | 277.18 | 554.37 | 1108.50 | 2217.73 | 4434.92 |
| **C** | 16.352 | 32.703 | 65.406 | 130.81 | 261.63 | 523.25 | 1046.50 | 2093.00 | 4186.01 |

# 120. HARTLEY OSCILLATOR

The frequency of a Hartley Oscillator determined by the LC resonant circuit, as shown in *Figure 111*.

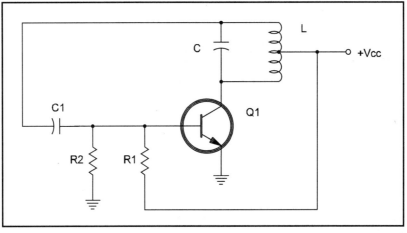

*Figure 111*

**Formula 120.1**
**Frequency**

$$f = \frac{1}{2 \times \pi \times \sqrt{L x C}}$$

Where:

F is the frequency in hertz (Hz)

L is the inductance in henry (H)

C is the capacitance in farads (F)

$\pi$ is 3.1416

# 121. COLPITTS OSCILLATOR

In the Colpitts Oscillator the feedback loop is a capacitive divider as shown in *Figure 112*. The frequency also depends on the LC circuit. The next formula is used to calculate.

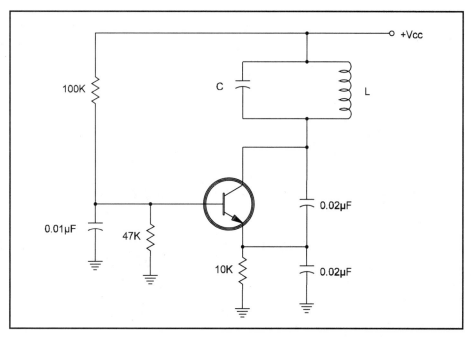

*Figure 112*

## Formula 121.1
## Frequency

$$f = \frac{1}{2 \times \pi \times \sqrt{L \times C}}$$

Where:

f is the frequency in hertz (Hz)

$\pi$ is the constant 3.1416

C is the capacitance in farads (F)

L is the inductance in henry (H)

**Application Example:**

Calculate the running frequency of a Colpitts Oscillator where the LC circuit is formed by a 100 pF capacitor and a 100 $\mu$ H coil.

Data:

L = 100 $\mu$ H

C = 100 pF

f = ?

Applying Formula 121.1:

$$f = \frac{1}{2 \times 3.14 \times \sqrt{100 \times 10^{-6} \times 100 \times 10^{-12}}} = \frac{1}{6.28 \times \sqrt{10^4 \times 10^{-6} \times 10^{-12}}}$$

$$f = \frac{1}{6.28 \times \sqrt{10^{-14}}} = \frac{10 \times 10^6}{6.28} = 1.592 \times 10^6 = 1.592 \text{ MHz}$$

# 122. CMOS TWO-GATE OSCILLATOR (I)

The simplest configuration for a two-gate oscillator or astable multivibrator using CMOS integrated circuits is shown in *Figure 113*. Any gate that can be wired as an inverter can be used in this application. The basic application recommends the use of a 4060 IC, but the 4001 and 4011 wired as inverters are also suitable.

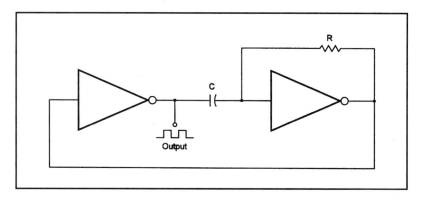

*Figure 113*

## Formula 122.1
## Period

$$T = -RxCx\left[\ln\frac{Vdd - Vtr}{Vdd}Vdd + \ln\frac{Vtr}{Vdd}\right]$$

Where:

T is the period in seconds (s)

R is the resistance in ohms ($\Omega$)

C is the capacitance in farads (F)

Vdd is the power supply voltage in volts (V)

Vtr is the transfer voltage in volts (V)

## Formula 122.2
## Reduced Formula to Calculate the Period

By letting Vtr = 0.5Vdd, formula 122.1 can be simplified:

$$T = 1.39 x R x C$$

Where:

T is the period in seconds (s)

R is the resistance in ohms ($\Omega$)

C is the capacitance in farads (F)

NOTE: 1.) In practice, Vtr can vary from 33% to 67% of Vdd.

2.) The output signal is a square wave with 50% of duty cycle.

## Formula 122.3
## Frequency (short form)

$$f = \frac{1}{1.39 x R x C}$$

Where:

f is the frequency in hertz (Hz)

R is the resistance in ohms ($\Omega$)

C is the capacitance in farads (F)

**Application Example:**

Calculate the frequency of a CMOS astable multivibrator with the configuration shown in *Figure 113*, where a 10 nF capacitor is used with a 100 kilohm resistor.

Data:

C = 10 nF = 10 x $10^{-9}$ F

R = 100 x $10^3$ $\Omega$

Using Formula 122.3:

$$f = \frac{1}{1.39x100x10^3 x10x10^{-9}} = \frac{1}{1.39x10^{-3}} = 0.719x10^3 = 719 \text{ Hz}$$

# 123.CMOS TWO-GATE OSCILLATOR (II)

The configuration shown in *Figure 113*, using two gates of a CMOS integrated circuit as inverters, has a problem: the maximum variation in the time period can be only as high as 9%. By adding a resistor (Rs) the frequency becomes independent from the supply voltage and the time-period variations with the power supply voltage is reduced.

The basic configuration of a two-gate CMOS oscillator or astable multivibrator with the described improvements is shown in *Figure 114*.

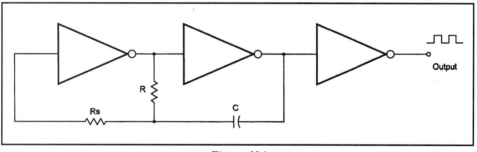

*Figure 114*

## Formula 123.1
## Exact Period

$$T = -RxCx\left[ \ln\frac{Vtr}{Vdd + Vtr} + \ln\frac{Vdd - Vtr}{2xVdd - Vtr} \right]$$

Where:

       T is the period in seconds (s)

       R is the resistance of R in ohms ($\Omega$)

       C is the capacitance in Farads (F)

       Vdd is the power supply voltage (V)

       Vtr is the transfer voltage (V)

NOTE: 1.) Rs should be 10 times the value of R.

2.) Minimum value recommended for R is 50 k$\Omega$

3.) C must be greater than 1 nF.

4.) Vtr in practice can vary from 33% to 67% of the power supply voltage (Vdd).

5.) The output is a square wave with 50% of duty cycle.

## Formula 123.2
## Simplified Formula (making Vtr =0.5xVdd)

$$T = 2.2xRxC$$

Where:

T is the period in seconds (s)

R is the resistance of R in ohms ($\Omega$)

C is the capacitance of C in farads (F)

## Formula 123.3
## Frequency (short form)

$$f = \frac{1}{2.2xRxC}$$

Where:

f is the frequency in hertz (Hz)

R is the resistance in ohms ($\Omega$)

C is the capacitance in farads (F)

**Application Example:**

In the circuit shown in *Figure 114*, R is a 10 kohm resistor, Rs a 100 kohm resistor and C a 0.05 $\mu$ F capacitor. Determine the frequency of the produced signal.

Data:

$$R = 10 \times 10^3 \; \Omega$$

$$C = 0.05 \times 10^{-6} \; F$$

$$f = ?$$

Using Formula123.3

$$f = \frac{1}{2.2x10x10^3x0.05x10^{-6}} = \frac{1}{1.1x10^{-3}} = 0.909x10^3 = 909 \text{ Hz}$$

## TABLE 34
## CMOS ICs Suitable for Using as Oscillators:

| Type | Function |
|------|----------|
| 4001 | Quad two-input NOR gate |
| 4011 | Quad two-input NAND gate |
| 4023 | Triple three-input NOR gate |
| 4025 | Triple three-input NAND gate |
| 4049 | Hex Inverter |
| 4069 | Hex Inverter |
| 4093 | Quad two-input NAND Schmitt triggers |

# 124. CMOS SCHMITT TRIGGER OSCILLATOR

The basic configuration of a 4093 wired as an astable multivibrator is shown in *Figure 115*. The output is in the ON state during the interval the capacitor charges from zero to the trigger point (T1). The output is LOW while the capacitor discharges from the trigger point to the holding point (T2). The output is a square wave with duty cycle of 50%.

The next formulas are valid for calculations involving this circuit:

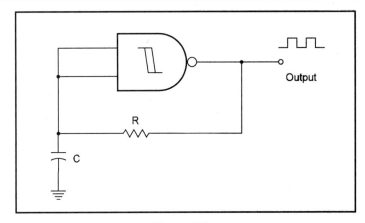

*Figure 115*

## Formula 124.1
## Periods

$$T1 = RxCx \ln\left(\frac{Vdd - Vn}{Vdd - Vp}\right)$$

$$T2 = RxCx \ln\left(\frac{Vp}{Vn}\right)$$

$$T = T1 + T2$$

Where:

T is the total period in seconds (s)

T1 is the period to the output in the high level in seconds (s)

T2 is the period to the output in the low level in seconds (s)

R is the resistance in ohms ($\Omega$)

C is the capacitance in farads (F)

Vdd is the power supply voltage in volts (V)

Vp is the positive threshold voltage in volts (V)

Vn is the negative threshold voltage in volts (V)

## TABLE 35
### Threshold Voltages of the 4093 CMOS IC

This table shows the values of threshold voltages as a function of the power supply voltage applied to the 4093 IC.

| Characteristic | Vdd | Value |
|---|---|---|
| Positive Threshold Voltage (Vp) | 5 | 1.8 |
| | 10 | 4.1 |
| | 15 | 6.3 |
| Negative Threshold Voltage (Vn) | 5 | 3.3 |
| | 10 | 6.2 |
| | 15 | 9.0 |

NOTE: The values in the table are typical for the 4093B IC from National Semiconductor. Small variations may be noted depending on the manufacturer. It is also important to note that these characteristics will change with temperature.

### Formula 124.2
### Frequency

$$f = \frac{1}{T}$$

$$f = \frac{1}{T1 + T2}$$

Where:

f is the frequency in hertz (Hz)

T is the total period given by Formula 124.1 in seconds (s)

T1 and T2 are the periods calculated by Formula 124.1 in seconds (s)

# 125. THE ASTABLE 555

The 555 integrated circuit can be used in two basic configurations: as an astable multivibrator producing square wave signals in a range of frequencies up to 500 kHz, and as a monostable multivibrator or timer giving time intervals up to one hour.

*Figure 116* shows the basic astable configuration. The following formulas are valid when used in calculations involving the 555 astable.

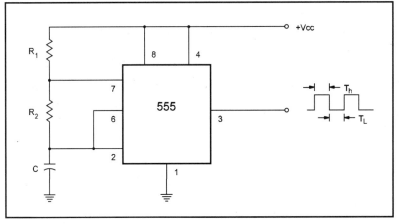

*Figure 116*

## Formula 125.1
## Charge Time of C (output high)

$$Th = 0.693xCx(R1 + R2))$$

Where:

Th is the time interval of the output high in seconds (s)

R1 and R2 are the resistances in ohms ($\Omega$)

C is the capacitance in farads (C)

## Formula 125.2
## Discharge Time of C (output low)

$$T_L = 0.693xR2xC$$

Where:

$T_L$ is the time interval of output low in seconds (s)

R2 is the resistance in ohms ($\Omega$)

C is the capacitance in farads (F)

## Formula 125.3
## Period

$$T = 0.693x(R1 + 2xR2)xC$$

Where:

T is the period in seconds (s)

R1 and R2 are the resistances in ohms ($\Omega$)

C is the capacitance in farads (F)

## Formula 125.4
## Frequency

$$f = \frac{1.44}{(R1 + 2xR2)xC}$$

Where:

f is the frequency in hertz (Hz)

R1 and R2 are the resistances in ohms ($\Omega$)

C is the capacitance in Farads (F).

## Formula 125.5
## Duty Cycle

$$Dc = \frac{Th}{T_L}$$

$$Dc = \frac{R1 + R2}{R2}$$

Where:

Dc is the duty cycle (0 to 1)

Th is the time the output is high in seconds (s)

TL is the time the output is low in seconds (s)

R1 and R2 are the resistances in ohms ($\Omega$)

## Formula 125.6
## Duty Cycle Percentage

$$Dc(\%) = Dcx100$$

Where:

Dc(%) is the duty cycle percentage (%)

Dc is the duty cycle (calculated by Formula 125.5)

## TABLE 36
## Limit Values Recommended For the Astable 555

| Component | Limit Value |
|---|---|
| R1 + R2 | max: 3 M$\Omega$ |
| R1 | min 1 k$\Omega$ |
| R2 | min 1 k$\Omega$ |
| C | min 500 pF |
| C | max 2 200 $\mu$ F (*) |
| f | max 1 MHz |
| Iout (drain or source) | 200 mA |
| Vcc | 18 V |

(*) Depends on the leakage

NOTE: There is a CMOS version of the bipolar 555—called TLC7555—with upgraded characteristics, including higher output current and upper frequency limit.

**Application Example:**

Determine the running frequency of a 555 in the astable configuration and using the following components: R1 = R2 = 10 kohm, and C1 = 50 nF.

Data:

R1 = R2 = 10 x $10^3$ Ω

C1 = 50 x $10^{-9}$ F

Using Formula 125.1:

$$f = \frac{1.44}{\left(10x10^3 x2x10x10^3\right)x50x10^{-9}} = \frac{1.44}{10^{-2}} = 1.44x10^2 = 144Hz$$

## GRAPH 1
## 555 Free Running Frequency

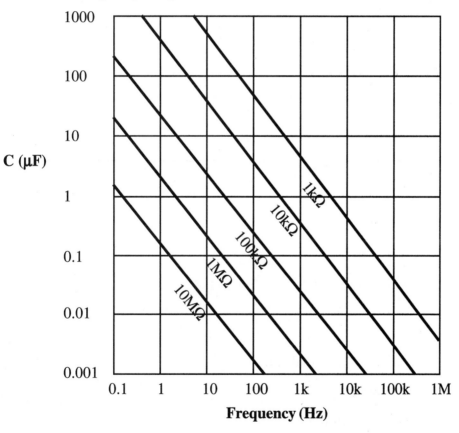

## 126. MONOSTABLE 555

When connected as a monostable multivibrator the 555 needs an external trigger command applied to the trigger input (pin 2) to start the action. This is normally done by leaving the trigger input positive and momentarily connecting it to to ground. At this the output goes to the high logic level (positive) by a time interval that can be calculated by the next formula. *Figure 117* shows the 555 in the monostable configuration.

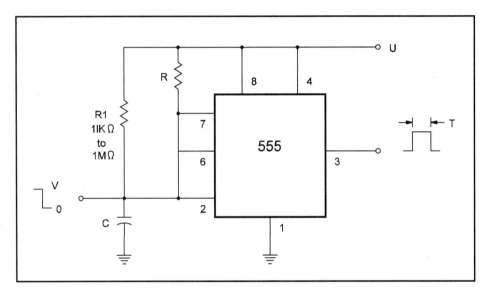

*Figure 117*

### Formula 126.1
### Time On

$$T = 1.1 x R x C$$

Where:

        T is the period or time ON in seconds (s)

        R is the resistance in ohms ($\Omega$)

        C is the capacitance in farads (F)

# TABLE 37

## Limit Values for the Monostable 555

| Parameter/Component | Limit Value |
| --- | --- |
| R max | 3 M$\Omega$ |
| R min | 1 k $\Omega$ |
| C max | 2 000 $\mu$F (*) |
| Cmin | 500 pF |
| Tr(max) | 1/4 T |
| Iout (drain or source) | 200 mA |
| Vcc | 18 V |

(*) depends on leakage.

Where:

Iout is the maximum output current

Vcc is the power supply voltage

Tr is the trigger pulse duration in seconds

**Application Example:**

Calculate the value of R, when with a 1000 $\mu$F, the ON time is 100 seconds using a 555 in the monostable configuration.

Data:

C = 1 000 x 10$^{-6}$ F

T = 100 s

R = ?

Using Formula 126.1:

$$100 = 1.1 x R x 1000 x 10^{-6}$$

Isolating R:

$$R = \frac{100}{1.1x1000x10^{-6}}$$

Solving the Equation:

$$R = \frac{100x10^3}{1.1} = 90.9x10^3 = 90.9 \text{ k}\Omega$$

## GRAPH 2
## 555 Monostable - On time versus R and C

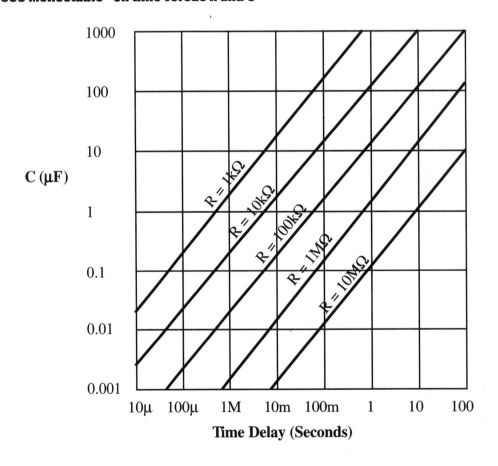

# BRIDGES

A bridge is a network formed by four arms connected and so arranged that when a signal (DC or AC) is applied across one pair of opposite junctions, the response of a detecting system (DC or AC indicator) connected between the other pair of junctions may be zeroed by adjusting one or more of the elements of the arms of the bridge.

The next formulas are applied to the most commonly used bridges, determining their balance conditions.

## 127. WHEATSTONE BRIDGE

This bridge is formed by resistors and used for resistance measurements. The circuit of a Wheatstone Bridge is shown in *Figure 118*. The detecting device can be a galvanometer. The applied signal is a DC voltage.

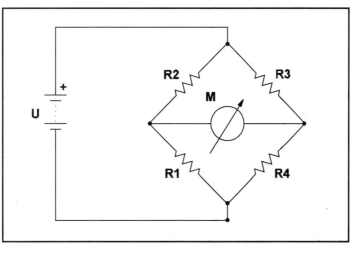

*Figure 118*

NOTE: 1.) When used to measure resistance, one of the resistances is unknown (R4 = Rx, for instance) and other is variable to balance the bridge (R3, for instance).

2.) The Wheatstone also operates from AC signal sources. You only have to use a compatible detecting device.

## Formula 127.1
### When Balanced

$$R4 = \frac{R1}{R2} x R3$$

Where:

R1, R2, R3, R4 are the resistances in the arms of the bridge in ohms ($\Omega$)

**Application Example:**

What is the value of R4 in the bridge shown in Figure 118, when the balance condition is found with R1 = 100 ohms, R2 =200 ohms and R3 = 300 ohms?

Data:

R1 = 100 ohms

R2 = 200 ohms

R3 = 300 ohms

R4 = ?

Using Formula 127.1:

$$R4 = \frac{100}{200} x 300 = 150 \text{ ohms}$$

# 128. WIEN BRIDGE

The Wien Bridge is used for capacitance and inductance measurements. *Figure 119* shows the basic configuration used in this bridge. The balance indicator is chosen according to the frequency of the used signal. The frequency is chosen according to the magnitude of the capacitance or inductance to be measured.

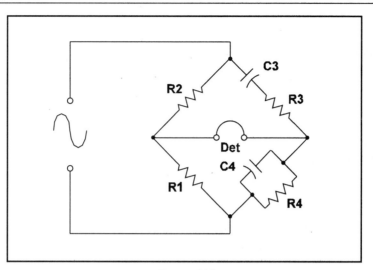

*Figure 119*

## Formula 128.1
## When Balanced

$$\omega^2 = \frac{1}{R4xR3xC4xC3}$$

*and*

$$\frac{C4}{C3} = \frac{R2}{R1} - \frac{R3}{R4}$$

Where:

    $\omega$ is 2 x $\pi$ x f  (f is the generator frequency in hertz - Hz)

    R1, R2, R3 and R4 are the resistance in ohms ($\Omega$)

    C1, C2, C3 and C4 are the capacitances in farads (F)

NOTE: The bridge can be used for frequency measurements. In this cases select the capacitors and resistor as follows:

    R3 = 2 x R4 and C1 = C2

Formula 128.2 can be used to determine the frequency of a signal source:

## Formula 128.2
## For frequency Measurements

$$f = \frac{1}{2 \times \pi \times R2 \times C2}$$

Where:

f is the frequency in hertz (Hz)

$\pi$ is the constant 3.1416

R2 is the resistance in ohms ($\Omega$)

C2 is the capacitance in farads (F)

## Formula 128.2
## When Balanced is also Valid

$$C4^2 = \frac{R2xR4 - R1xR3}{R1xR4^2 \, xR3x\omega^2}$$

*and*

$$C3^2 = \frac{R1}{(R2xR4 - R1xR3)xR3x\omega^2}$$

Where:

$\omega = 2 \times \pi \times f$ (f is the frequency in hertz - Hz)

R1, R2, R3 and R4 are the resistances in ohms ($\Omega$)

C1, C2, C3 and C4 are the capacitances in farads (F)

# 129. RESONANCE BRIDGE

The Resonance Bridge is used for inductance measurements and needs a signal source with frequency determined by the magnitude of the measured quantities. The circuit is shown in *Figure 120* and there are two elements to be adjusted for balance: C and R3.

C4 and R4 are the capacitance and resisitance representing the inductance to be measured.

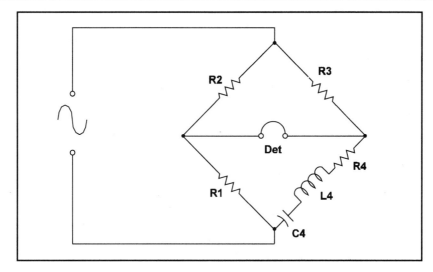

*Figure 120*

## Formula 129.1
## Condition to Balance

$$L = \frac{1}{\omega^2 \, xC}$$

*and*

$$R4 = \frac{R1}{R2} \, xR3$$

Where:

$\omega = 2 \times \pi \times f$  (f is the frequency in hertz - Hz)

R1, R2, R3 and R4 are the resistances in ohms ($\Omega$)

C is the capacitance in farads (F)

L is the inductance in henrys (H)

# 130. MAXWELL BRIDGE

This bridge is intended for inductance measurements and the standard is a capacitance. This makes it ideal in practical applications, since when using inductances as standard the costs are higher. The basic circuit of a Maxwell Bridge is shown in *Figure 121*. The bridge uses a signal source with frequency determined by the magnitude of the measured inductance. Notice that the balance point is independent of the frequency.

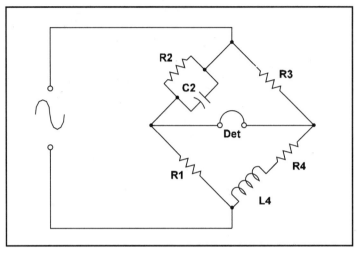

*Figure 121*

## Formula 130.1
## When Balanced

$$L4 = R1 x R2 x C$$

*and*

$$R4 = \frac{R1}{R2} x R3$$

Where:

L4 is the inductance to be measured in henrys (H)

R1, R2, R4 and R4 are the resistance in the arms in ohms ($\Omega$)

C is the capacitance in farads (F)

**Application Example:**

In the Maxwell Bridge shown in *Figure 121* the balance is found when: R1 = R2 = 100 ohms, R3 = 200 ohms and C = 0.1 $\mu$ F. Determine L and R4. (R4 is resistance of the coil winding).

Data:

> R1 = R2 = 100 ohms
>
> R3 = 200 ohms
>
> C = 0.1 $\mu$ F (when using C in $\mu$ F the result is found in $\mu$ H)

Using Formula 130.1:

$$L4 = 100x200x0.1 = 2000\mu H = 2mH$$

$$R4 = \frac{100}{100}x200 = 200ohms$$

# 131. SCHERING BRIDGE

This bridge is used to measure inductances and capacitances. The input is a signal generator with frequency chosen according to the magnitude of the capacitance or inductance to be measured. The basic circuit for capacitance measurements is shown in *Figure 122*.

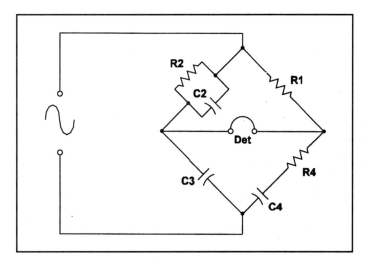

*Figure 122*

## Formula 131.1
## When Balanced

$$C4 = \frac{R2}{R1} x C3$$

*and*

$$R4 = \frac{C2}{C3} x R1$$

Where:

C2 and C3 are the capacitances in the arms (one of them variable) in farads (F)

C4 is the unknown capacitance in farads (F)

R4 is the unknown resistance in ohms ($\Omega$)

R1 and R2 are the resistances in the arms (one of them variable) in ohms ($\Omega$)

**Application Example:**

A Schering Bridge is balanced when R1 = R2 = 200 ohms and when C1 = C3 = 0.1 $\mu$ F. Determine the unknown capacitance (C4) and the associated resistance (R4).

Data:

R1 = R2 = 200 ohms

C1 = C3 = 0.1 $\mu$ F

Determining C4:

$$C4 = \frac{200}{200} x 0.1 = 0.1 \mu F$$

Determining R4:

$$R4 = \frac{0.1}{0.1} x 200 = 200 ohms$$

# 132. OWEN BRIDGE

This bridge is used for inductance measurements and the results are independent of the generator frequency. Despite this, the frequency of the external generator can be chosen according to the magnitude of the measured inductance. The basic circuit is shown in Figure 123.

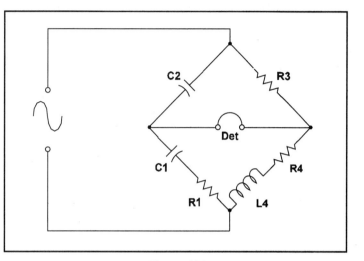

*Figure 123*

## Formula 132.1
## When Balanced

$$L4 = R1xR3xC2$$

*and*

$$R4 = \frac{C2}{C1}xR3$$

Where:

    L4 is the unknown inductance in henrys (H)

    R1 and R3 are the resistances in the arms in ohms ($\Omega$) - R1 variable

    R4 is the resistance associated to the inductance in ohms ($\Omega$)

    C1, C2 and C3 are capacitances in farads (F) - one of them variable (C1)

# 133. HAY BRIDGE

This bridge is used for inductance measurements and the results are dependent from the signal generator frequency. So, the frequency must be chosen according to the magnitude of the inductance to be measured. The basic circuit is shown in *Figure 124*.

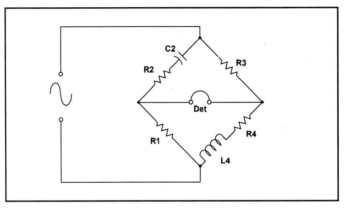

*Figure 124*

## Formula 124.1
## When Balanced

$$L = \frac{R1 x R2 \, XC2}{1 + (R2 x \omega x C2)^2}$$

*and*

$$R4 = \frac{R1 x R2 x R3(\omega x C2)^2}{1 + (R2 x \omega x C2)^2}$$

Where:

$\omega = 2 x \pi x f$ (f is the frequency in hertz - Hz)

L is the unknown inductance in henrys (H)

R1, R2, R3 are resistances in ohms ($\Omega$) - two of them variable to balance the bridge.

R4 is the resistance associated with the coil in ohms ($\Omega$)

C1 and C2 are capacitances in farads (F)

# OPERATIONAL AMPLIFIERS

Although operational amplifiers were originally used to perform mathematical operations on electrical analog systems or either physical phenomena, their applications are limited only by the imagination and technical acumen of the designer. Next are formulas suitable to projects involving operational amplifier configurations. The reader must remember that in real life operational amplifiers are not ideal components and can present a wide range of characteristics. This means that the practical circuits can have different performance when compared to the calculations.

## 134. NONINVERTING OP AMP

In the noninverting configuration the phase of the input signal is the same as the output signal. The gain is determined by the feedback resistance. The basic circuit of a noninverting operational amplifier application is shown in *Figure 125*. The following formulas is used to determine the gain.

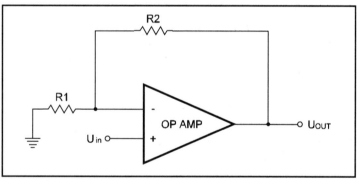

*Figure 125*

**Formula 134.1**
**Gain**

$$G = 1 + \frac{R2}{R1}$$

Where:

G is the gain

R1 and R2 are the resistances in ohms ($\Omega$)

## Formula 134.2
## Voltage Gain

$$Uout = Uinx\left(1 + \frac{R2}{R1}\right)$$

Where:

Uout is the output voltage in volts (V)

Uin is the input voltage in volts (V)

R1 and R2 are the resistances in ohms ($\Omega$)

NOTE: The output voltage cannot exceed the power supply voltage.

**Application Example:**

In the circuit shown in *Figure 125*, R2 is a 100 kohm resistor and R1 is a 10 kohm resistor. Determine the gain.

Data:

R1 = 10 k$\Omega$

R2 = 100 k$\Omega$

G = ?

Applying Formula 134.1

$$G = 1 + \frac{100x10^3}{10x10^3} = 1 + 10 = 11$$

# 135. INVERTING OP AMP

The polarity of the output signal is opposite the input signal. The basic configuration of an inverting operational amplifier application is shown in *Figure 126*.

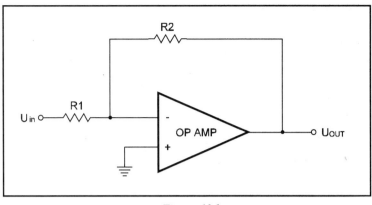

*Figure 126*

**Formula 135.1**

**Gain of an Inverting Amplifier**

$$G = -\frac{R2}{R1}$$

Where:

G is the gain

R1 and R2 are the resistances in ohms ($\Omega$)

# VOLTAGE FOLLOWER

The voltage follower is a special configuration where both R1 and R2 are zero. This configuration has a unity gain as shown in *Figure 127*.

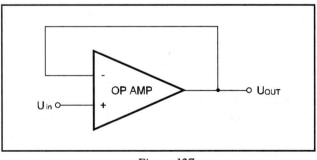

*Figure 127*

# 136. SUMMING OP AMP

The basic configuration of a summing amplifier using an operational amplifier is shown in *Figure 128*. The output voltage is the algebraic sum of the input voltages times the gain given by the ratio of R2 and R1.

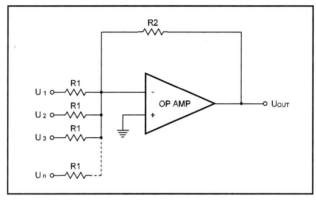

*Figure 128*

## Formula 136.1
## Summing Amplifier

$$Uout = \frac{R2}{R1} x (U1 + U2 + U3)$$

Where:

      Uout is the output voltage in volts (V)

      U1, U2....Un are the input voltages in volts (V)

      R1, R2 are the resistances in ohms ($\Omega$)

**Application Example:**

In the summing amplifier shown in *Figure 129*, R2 is a 100 kohm resistor and R1 is 10 kohm. Determine the output voltage when the input voltages are U1 = 100 mV, U2 = -200 mV and U3 = 250 mV.

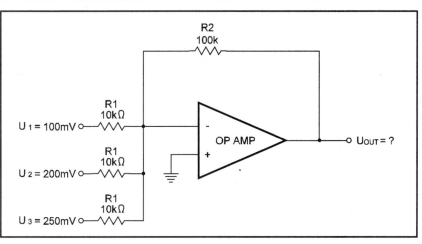

*Figure 129*

Data:

R2 = 100 k$\Omega$

R1 = 10 k$\Omega$

U1 = 100 mV

U2 = -200 mV

U3 = 250 mV

Uout = ?

Applying Formula 136.1:

$$Uout = \frac{100x10^3}{10x10^3} \, x(100x10^{-3})$$

$$Uout = 10x(150x10^3) = 150 \quad mV = 1.5 \text{ V}$$

# 137. SUBTRACTION OP AMP

The output voltage is the difference between the input voltages times the gain. This is the aim of the configuration shown in *Figure 130*.

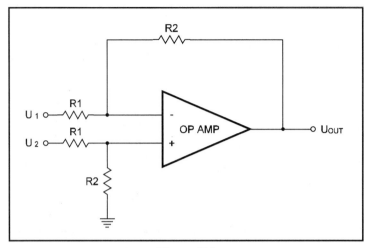

*Figure 130*

## Formula 137.1
## Subtraction

$$Uout = \frac{R2}{R1} x (U2 - U1)$$

Where:

Uout is the output voltage in volts (V)

U1 and U2 are the input voltages in volts (V)

R1 and R2 are the resistances in ohms ( $\Omega$ )

NOTE: The output voltage must be less than the power supply voltage.

## TABLE 38
## CMRR vs dB

CMRR means Common Mode Rejection Ratio and is an important parameter for projects involving operational amplifiers. The CMRR defines the ability of any particular device to reject common-mode input voltages. The next table gives the common values of CMRR converted to dB.

| CMRR | dB |
|---|---|
| 1 | 0 |
| 10 | 20 |
| 100 | 40 |
| 1 000 | 60 |
| 10 000 | 80 |
| 100 000 | 100 |
| 1 000 000 | 120 |
| 10 000 000 | 140 |

# 138. DIFFERENTIATION USING OP AMP

Differentiation is a process used to find the instantaneous rate of change of a signal by finding the slope of a line tangent to the point of interest on the graph of the function of the signal. *Figure 131* shows a basic differentiation circuit using an operational amplifier.

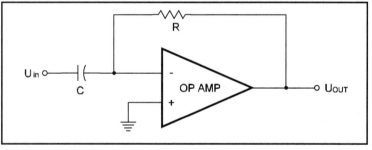

*Figure 131*

**Formula 138.1**
**Differentiation**

$$Uout = -RxCx\left(\frac{dUin}{dt}\right)$$

Where:

Uout is the instantaneous output voltage in volts (V)

Uin is the input voltage in volts (V)

R is the resistance in ohms ($\Omega$)

C is the capacitance in farads (F)

# 139. INTEGRATION USING OP AMP

*Figure 132* shows how can an operational amplifier can be used to perform the integration function. The switch is used to determine the starting point of the process as the output voltage is a function of time.

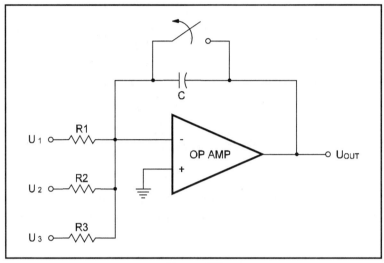

*Figure 132*

## Formula 139.1
## Integration

$$Uout = \frac{1}{C}\int(\frac{U1}{R1}+\frac{U2}{R2}+\frac{U3}{R3})$$

Where:

Uout is the output voltage in volts (V)

U1, U2, U3....Un are the input voltages in volts (V)

R1, R2, R3....Rn are the resistances in ohms ($\Omega$)

C is the capacitance in farads (F)

# 140. LOGARITHMIC OP AMP

In a logarithmic amplifier the gain depends on the input voltage and input current. The circuit shown in *Figure 133* uses a transistor as a variable feedback device, determining the voltage gain as a function of the input voltage and input current. The next formulas can be applied in calculations involving this circuit.

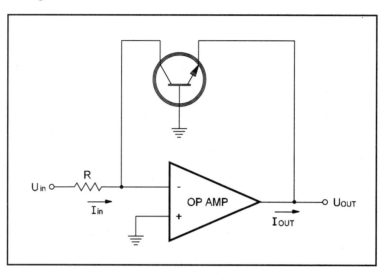

*Figure 133*

## Formula 140.1
## Voltage Gain

$$Uout = -Uinx \lg(\frac{Iin}{Iout})$$

Where:

Uout is the output voltage in volts (V)

Uin is the input voltage in volts (V)

Iin is the input current in amperes (A)

Iout is the output current in amperes (A)

### Formula 140.2
### Input Current

$$Iin = \frac{Uin}{R}$$

Where:

Iin is the input current in amperes (A)

Uin is the input voltage in volts (V)

R is the resistance in ohms ($\Omega$)

NOTE: The output voltage cannot exceed the power supply voltage. For common operational amplifiers 15V is a typical limit.

# 141. VOLTAGE SOURCE

This configuration can be used as a reference voltage source, and, if the operational amplifier used is a high-power type, as a high power zener. The circuit is shown in *Figure 134* and the formula given below can be used in practical applications.

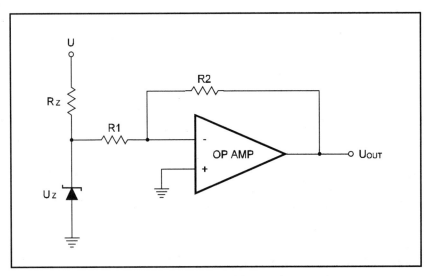

*Figure 134*

# Formula 141.1
# Voltage Source

$$Uout = -Uzx\left(\frac{R2}{R1}\right)$$

Where:

Uout is the output voltage in volts (V)

Uz is the zener voltage

R1 and R2 are the resistances in ohms ($\Omega$)

NOTE: Rz is calculated according to the formulas described in the Zener Diode section (formulas 87.x).

**Application Example:**

Determine the value for R2 when desiring a reference output voltage of 7V from a 3V zener diode in the circuit shown in *Figure 134*. Given R1= 500 ohms.

Data:

Uz = 3 V

Uout = 7 V

R1 = 500 ohms

R2 = ?

Using Formula 141.1:

$$7 = 3x\left(\frac{R2}{500}\right)$$
$$7 = \frac{3xR2}{500}$$

Isolating R2:

$$R2 = \frac{7x500}{3}$$

$$R2 = \frac{3500}{3} = 1166 ohms$$

# 142. CONSTANT CURRENT SOURCE - Op Amp (Floating Load)

The circuit shown in *Figure 135* is indicated for floating loads. The resistor R in series with the zener diode is calculated using the formulas indicated in section 87.x.

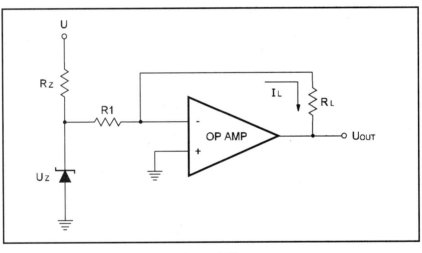

*Figure 135*

## Formula 142.1
## Constant Current Source - Floating Load

$$I_L = \frac{Uz}{R1}$$

Where:

$I_L$ is the load current in amperes (A)

Uz is the zener voltage in volts (V)

R1 is the resistance in ohms ($\Omega$)

NOTE: Rz is calculated according to the zener formulas (formulas 87.x).

# 143. CONSTANT CURRENT SOURCE - Op Amp (High Current Load)

Another configuration suitable to large current loads is shown in *Figure 136*. The reference voltage is applied to the input and can be sourced to the load by a voltage source using another operational amplifier.

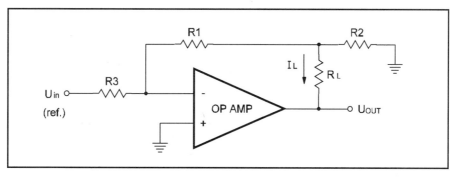

*Figure 136*

## Formula 143.1
## Constant Current Source - High Power

$$I_L = Uinx\left[\frac{R1x(R1+R2)}{R1xR2xR3}\right]$$

Where:

$I_L$ is the current in the load in amperes (A)

Uin is the input voltage or reference voltage in volts (V)

R1, R2 and R3 are the resistances in ohms ($\Omega$)

# 144. ABSOLUTE VALUE AMPLIFIER - Op Amp

In this circuit the voltage at the output is the absolute value of the input voltage times the gain. The gain depends on R2 and R1, according the following formulas. The circuit is shown in *Figure 137*.

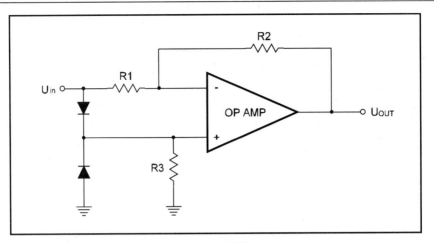

*Figure 137*

## Formula 144.1
## Absolute Value Amplifier

$$G = \frac{R2}{R1}$$

*and*

$$Uout = |Uin| x \left( \frac{U2}{U1} \right)$$

Where:

G is the gain

Uout is the output voltage in volts (V)

Uin is the input voltage in volts (V)

R1 and R2 are the resistances in ohms ($\Omega$)

NOTE: 1.) Uout must be lower than the power supply voltage.

2.) R3 is chosen according the application. Typical values are R3 = R1.

# 145. VOLTMETER USING Op Amp

A high-impedance input and low-voltage voltmeter is shown in *Figure 138*.

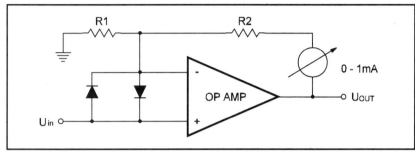

*Figure 138*

## Formula 145.1
## High-Impedance Voltmeter

$$\mathrm{Im} = \frac{Uin}{R1}$$

Where:

Im is the current flowing through the meter in amperes (A)

Uin is the input voltage in volts (V)

R1 is the resistance in ohms ($\Omega$)

NOTE: R2 is the current limiting resistor and is chosen according to the meter.

# 146. SQUARE WAVE OSCILLATOR - Op Amp

The relaxation circuit shown in *Figure 139* using an operational amplifier can produce signals from very low frequencies (under 1 Hz) up to 100 kHz depending on the op amp used. The output is a square wave, but in the junction between R and C the signal is a sawtooth.

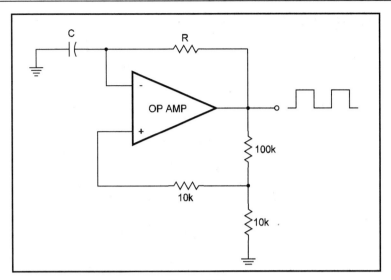

*Figure 139*

## Formula 146.1
## Square Wave Oscillator

$$f = \frac{1}{6xRxC}$$

Where:

f is the frequency in hertz (Hz)

R is the resistance in ohms ($\Omega$)

C is the capacitance in farads (F)

**Application Example:**

Determine C to a frequency of 10 kHz given R = 10 kohm. The circuit is shown in *Figure 139*.

Data:

f = 10 kHz

R = 10 k$\Omega$

C = ?

Using Formula 146.1:

$$10x10^3 = \frac{1}{6x10x10^3 xC}$$

*isolatingC*:

$$C = \frac{1}{6x10x10^3 x10x10^3}$$

$$C = \frac{1}{600x10^6} = \frac{1}{600}x10^{-6}$$

$$C = 0.0016uF$$

$$C = 1.6nF$$

# 147. WIEN BRIDGE OSCILLATOR using Op Amp

A Wien Bridge oscillator using an operational amplifier is shown in *Figure 140*. The zener diode is used for stabilization of the sine wave output.

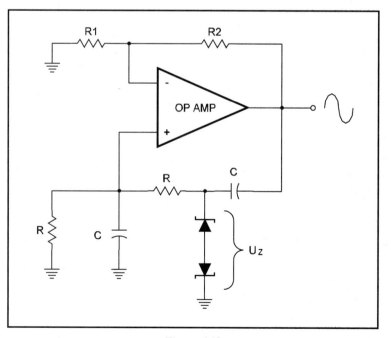

*Figure 140*

## Formula 147.1
## Wien Bridge Oscillator

$$f = \frac{1}{2 \times \pi \times R \times C}$$

Where:

f is the frequency in hertz (Hz)

R is the resistance in ohms ( $\Omega$ )

C is the capacitance in farads (F)

$\pi$ is the constant 3.1416

# 148. BANDPASS AMPLIFIER using Op Amp

The bandpass filter shown in *Figure 141* can use in the feedback loop of an LC resonant circuit or RC network. The characteristics (bandwidth, Q-factor and others) are determined by the type of filter. The next formulas are used for calculations involving output frequency.

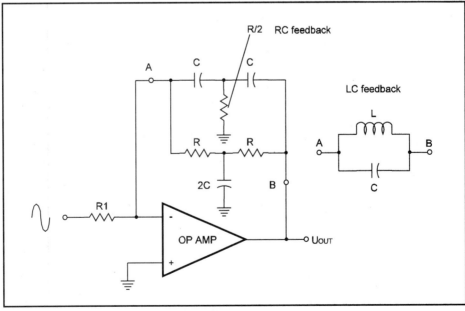

*Figure 141*

# Formula 148.1
## LC Feedback

$$f = \frac{1}{2 \times \pi \times \sqrt{L \times C}}$$

Where:

    f is the frequency in hertz (Hz)

    L is the inductance in henry (H)

    C is the capacitance in farads (F)

    $\pi$ is 3.1416

# Fomula 148.2
## RC Feedback Network

$$f = \frac{\sqrt{3}}{2 \times \pi \times R \times C}$$

Where:

    f is the frequency in hertz (Hz)

    R is the resistance in ohms ($\Omega$)

    C is the capacitance in farads (F)

    $\pi$ is 3.1416

    $\sqrt{3} = 1.7320$

**Application Example:**

Determine R to a bandpass filter using the configuration shown in *Figure 141* tuned to 10 kHz where the used capacitors are 0.005 $\mu$ F units (5 nF).

Data:

    f = 10 kHz

    C = 0.005 x $10^{-6}$

    R = ?

Using Formula 148.2:

$$10x10^3 = \frac{1.73}{2x3.14xRx0.005x1}$$

$$R = \frac{1.73}{6.28x0.005x10^{-6}x10x10}$$

$$R = \frac{1.73}{0.628x10^{-3}} = 2.754x10$$

$$R = 2.754 \text{ k}\Omega$$

# 149. NOTCH FILTER - Op Amp

The filter shown in *Figure 142* uses an RC feedback network and the signals of the tuned frequency are rejected.

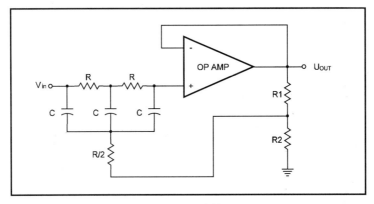

*Figure 142*

**Formula 149.1**
**Notch Filter Frequency**

$$f = \frac{\sqrt{3}}{2x\pi xRxC}$$

Where:

f is the frequency in hertz (Hz)

R is the resistance in ohms ($\Omega$)

C is the capacitance in farads (F)

$\pi$ is 3.1416

$\sqrt{3} = 1.7320$

# 150. LOW-PASS FILTER - Op Amp

The low-pass filter shown in *Figure 143* uses two RC networks. Frequencies under the calculated value pass without attenuation. Frequencies above the calculated value are blocked.

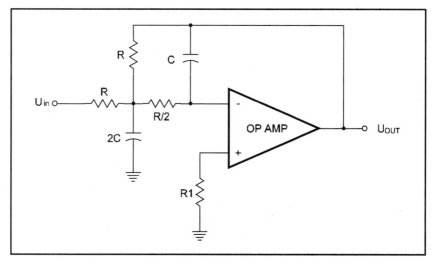

*Figure 143*

**Formula 150.1**
**Low-Pass Filter**

$$f = \frac{1}{2 \times \pi \times R \times C}$$

Where:

    f is the frequency in hertz (Hz)

    R is the resistance in ohms ($\Omega$)

    C is the capacitance in farads (F)

    $\pi = 3.1416$

    R3 = 2 x R

# 151. HIGH-PASS FILTER - Op Amp

The circuit shown in *Figure 144* uses only resistors and capacitors. Frequencies up to the calculated value pass without attenuation. Frequencies under the calculated value are blocked.

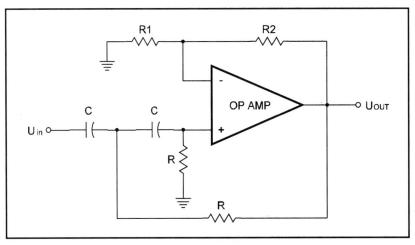

*Figure 144*

**Formula 151.1**
**High-Pass Filter**

$$f = \frac{1}{2 \text{x} \pi \text{x} R \text{x} C}$$

Where:

f is the frequency in hertz (Hz)

R is the resistance in ohms ($\Omega$)

C is the capacitance in farads (F)

$\pi$ is 3.1416

# 152. BUTTERWORTH FILTER - Op Amp

The circuit shown in Figure 145 is a second-order Butterworth filter using an operational amplifier. The next formulas can be used for calculations involving its performance.

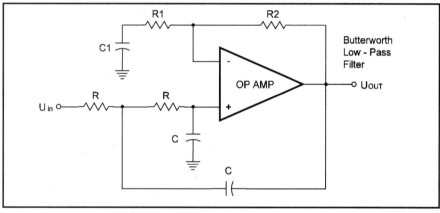

*Figure 145*

## Formula 152.1
## Frequency

$$f = \frac{1}{2 \times \pi \times R \times C}$$

Where:

f is the frequency in hertz (Hz)

R is the resistance in ohms ($\Omega$)

C is the capacitance in farads (F)

$\pi$ is 3.1416

**Formula 152.2**
**Selectivity**

$$Q = \frac{3}{3-G}$$

Where:

Q is the Q-factor

G is the voltage gain (see next formula)

**Formula 152.3**
**Gain**

$$G = 1 + \frac{R2}{R1}$$

Where:

G is the gain

R1 and R2 are the resistances in ohms ($\Omega$)

# 153. SERIES VOLTAGE REGULATOR (one transistor)

The configuration shown in *Figure 146* is the basic voltage regulator using a zener diode as reference and a transistor to control the load current. The following formulas are used when calculating the elements in this circuit:

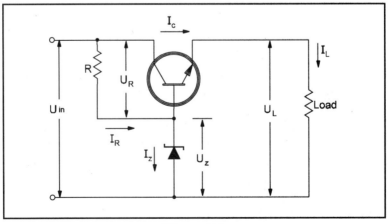

*Figure 146*

## Starting Point:

In a practical project it is important to start from some fixed parameters and calculate the used components from them. The fixed parameters to be used in the next formulas are:

- $U_L$ = voltage in the output (load) in volts (V)

- Uin(min) = minimum input voltage in volts (V)

- Uin(max) = maximum input voltage in volts (V)

- Uin = input voltage (value between Uin(max) and Uin(min) - normally the variation of the input voltage adopted in the projects are 10%) in volts (V)

- Iz(min) = minimum current through the zener (indicated by manufacturer) in amperes (A)

- $I_{B(max)}$ = maximum value of the base current (indicated by the manufacturer or chosen according to the transistor gain as function of the load current) in amperes (A)

**Important: Uin > Uz**

## Formula 153.1
## Load Voltage

$$U_L = Uz + U_{BE}$$

Where:

UL is the load voltage in volts (V)

Uz is the zener voltage in volts (V)

$U_{BE}$ is the voltage fall in the base-emitter junction (typically 0.6V for silicon transistors) in volts (V)

## Formula 153.2
## Voltage Across R

$$U_R = Uin - Uz$$

Where:

$U_R$ is the voltage across R in volts (V)

Uin is the input voltage in volts (V)

Uz is the zener voltage in volts (V)

NOTE: The input voltage must be greater than the zener voltage for correct operation. A difference at least of 3 volts is recommended in normal projects.

## Formula 153.3
## Iz(max)

$$Iz(max) = \left[ \frac{Uin(mx) - Uz}{Uin(min) - Uz} \right]$$

Where:

Iz(max) is maximum current in the zener diode in amperes (A)

Uin(max) is the maximum input voltage in volts (V)

Uin(min) is the minimum input voltage in volts (V)

Uz is the zener voltage in volts (V)

Iz(min) is the minimum current in the zener in amperes (A)

IB(max) is the maximum current in the transistor's base in volts (V)

## Formula 153.4
## IB(max)

$$Iz(max) = \frac{Ic(max)}{\beta(min)}$$

Where:

Iz(max) is the maximum current in the zener diode in amperes (A)

Ic(max) is maximum current in the transistor collector (load current) in amperes (A)

$\beta(min)$ is the minimum gain of the used transistor

# Formula 153.5
# Zener Dissipation

$$Pz = Iz(max)xUz$$

Where:

Pz is the minimum power dissipation of the used zener in watts (W)

Iz(max) is the maximum current in the zener diode in amperes (A)

Uz is the zener voltage in volts (V)

# Formula 153.6
# Calculating R

$$Rmin = \frac{Uin(max) - Uz}{Iz(max)}$$

*and*

$$Rmax = \frac{Uin(min) - Uz}{IB(max) + Iz(min)}$$

$$Rmin < R < Rmax$$

Where:

R is the chosen resistor in ohms ( $\Omega$ )

Rmin is the minimum value for R in ohms ( $\Omega$ )

Rmax is the maximum value for R in ohms ( $\Omega$ )

Uin(max) is the maximum input voltage in volts (V)

Uin(min) is the minimum input voltage in volts (V)

Uz is the zener voltage in volts (V)

Iz(max) is the maximum current in the zener diode in amperes (A)

Iz (min) is the minimum current in the zener diode in amperes (A)

IB(max) is the maximum base current in the transistor in amperes (A)

# Formula 153.7
## Power Dissipation of R

$$P = \frac{U_R^2}{R}$$

Where:

P is the power dissipated by R in watts (W)

$U_R$ is the voltage across R in volts (V)

R is the resistance in ohms ($\Omega$)

# Formula 153.8
## Minimum Input Voltage

$$Uin(min) = Uz + R(Iz(min)$$

Where:

Uin(min) is the minimum input voltage in volts (V)

Uz  is the zener voltage in volts (V)

R is the resistance in ohms ($\Omega$)

Iz(min) is the minimum current in the zener in amperes (A)

Ib(max) is the maximum current in the transistor base in amperes (A)

# Formula 153.9
## Determining Minimum Gain for the Transistor

$$\beta(min) = \frac{Ic(max)}{I_B(max)}$$

Where:

$\beta$ (min) is the minimum gain of the transistor used

Ic(max) is the maximum current in the transistor (load current) in amperes (A)

$I_B$(max) is the maximum base current in the transistor in amperes (A)

## Application Example:

Calculate the practical elements of a voltage regulator with configuration shown in *Figure 147*.

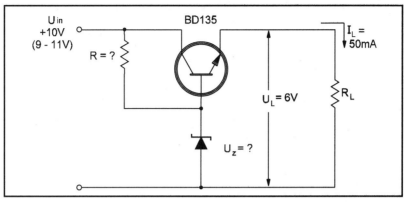

*Figure 147*

Data:

Uin = 10V (10% of variation)

Uin(max) = 11V

Uin(min) = 9V

UL= 6V

IL = 500 mA

Zener parameters will be fixed after the voltage is determined.

Calculating Uz using Formula 153.1:

Uz = Uin + 0.6 (we intend to use a silicon transistor)

Uz = 10 - 3.4

Uz = 6.6V (we adopted the voltage average input value)

Calculating $U_R$

$$U_R = Uin - Uz$$
$$U_R = 10 - 6.6$$

$$U_R = 3.4 \text{ V}$$

Determining Iz(max) using Formula 153.4:

$$Iz(max) = \left[\frac{Uin(max) - Uz}{Uin(min) - Uz}\right] x\big(Iz(min) + I_B(max)\big)$$

$$Iz(max) = \left[\frac{11 - 6.6}{9 - 6.6}\right] x(0.01 + 002)$$

$$Iz(max) = \frac{4.4}{2.4} x0.03 = 0.054A$$

Calculating R using Formula 153.6:

$$Rmin = \frac{11 - 6.6}{0.054}$$

$$Rmin = \frac{4.4}{0.054} = 81.4\Omega$$

$$Rmax = \frac{9 - 6.6}{0.02 + 0.01}$$

$$Rmax = \frac{3.4}{0.03} = 113\Omega$$

$81.4 < R < 113$ ohms

Use a 100 ohm resistor for R.

Determining the minimum $\beta$ for the transistor, using Formula 153.9:

$$\beta(min) = \frac{0.5}{0.02} = 25$$

Determining the power dissipation of R, using Formula 153.7:

$$P_R = \frac{(3.4)^2}{100} = 0.11W$$

The used $U_R$ is for Uin(max).

A 1/4W or 1/2W resistor can be used in the practical circuit.

Calculating the dissipation for the zener diode using Formula 153.5:

$$Pz = 6.6x0.054 = 0.356W$$

A 1/2W or 1W zener diode can be used in practice.

## TABLE 39

## Common Values of Expressions using $\pi$

| | | | |
|---|---|---|---|
| $\pi$ | 3.141592 | $\pi/5$ | 0.628318 |
| $2\pi$ | 6.283185 | $4\pi/3$ | 4.188790 |
| $3\pi$ | 9.424779 | $4/\pi$ | 1.273239 |
| $4\pi$ | 12.566379 | $3/\pi$ | 0.954929 |
| $5\pi$ | 15.707963 | $\pi^2$ | 9.869604 |
| $6\pi$ | 18.849556 | $\sqrt{\pi}$ | 1.772453 |
| $7\pi$ | 21.991486 | $\dfrac{1}{\pi^2}$ | 0.101321 |
| $8\pi$ | 25.132741 | | |
| $9\pi$ | 28.274334 | $\dfrac{1}{\sqrt{\pi}}$ | 0.564190 |
| $10\pi$ | 31.415926 | | |
| $1/\pi$ | 0.318310 | $\sqrt{\dfrac{3}{\pi}}$ | 0.977205 |
| $1/2\pi$ | 0.159155 | | |
| $\pi/2$ | 1.570796 | $3\sqrt{\pi}$ | 1.464592 |
| $\pi/3$ | 1.047197 | $\lg\pi$ | 0.497715 |
| $\pi/4$ | 0.785398 | $\lg2\pi$ | 0.798180 |

# 154. PARALLEL VOLTAGE REGULATOR

The basic configuration shown in *Figure 148* is a parallel regulator or shunt regulator. The transistor is wired in parallel with the load, shunting the current in a way so the load voltage remains constant.

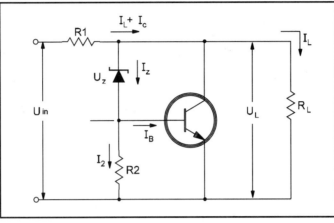

*Figure 148*

The elements to be calculated are the zener charateristics, R1 and R2 and also the transistor characteristics. The next formulas are used when some parameter are fixed.

The fixed parameters are:

- $U_L$ = load voltage in volts (V)

- $I_L$ (max) = maximum current in the load in amperes (A)

- Ic(min) = minimum current flowing by the collector of Q in amperes (A)

- Iz(min) = minimum current flowing through the zener diode in amperes (A)

- Uin = input voltage in volts (V)

- Uin(max) = maximum input voltage (normally 10% over Uin)

- Uin(min) = minimum input voltage (normally 10% under Uin)

- $U_{BE}$ = voltage between base end emitter - 0.6 V for silicon transistors

- $\beta$ (min) = minimum gain of the used transistor

## Formula 154.1
## Determining Uz

$$Uz = U_L - 0.6$$

Where:

Uz is the zener voltage in volts (V)

$U_L$ is the voltage across the load in volts (V)

0.6 is a constant when using a silicon transistor

## Formula 154.2
## $I_{R2}$ (current through R2)

$$I_{R2} = Iz(min) - \frac{Ic(min)}{\beta min}$$

Where:

$I_{R2}$ is the current through R2 in amperes (A)

Iz(min) is the minimum current through the zener diode in amperes (A)

Ic(min) is the minimum current through the transistor in amperes (A)

$\beta min$ is the minimum gain of the transistor

## Formula 154.3
## Determining Iz(max)

$$Iz(max) = \left[ \left( \frac{Uin(max - Uz \cdot}{Uin(min) - Uz} \right. \right.$$

Where:

Iz(max) is the maximum current flowing through the zener (A)

Uin(max) is the maximum input voltage (V)

Uin(min) is the minimum input voltage (V)

(continued)

Uz is the zener voltage in volts (V)

Iz(min) is the minimum zener current in amperes (A)

Iz(max) is the maximum zener current in amperes (A)

Ic(min) is the minimum collector current in amperes (A)

*βmin* is the minimum transistor gain

$I_{R2}$ is the current through R2 in amperes (A)

## Formula 154.4
## Calculating Ic(max)

$$Ic(max) = \beta minx\left(Iz(max) - I_{R2}\right)$$

Where:

Ic(max) is the maximum current in the transistor in amperes (A)

*βmin* is the minimum transistor gain

Iz(max) is the maximum zener current in amperes (A)

$I_{R2}$ is the current through R2 in amperes (A)

## Formula 154.5
## Pc(max) - Power Dissipated by the Transistor

$$P(max) = (Uz + 0.6)xIc(max)$$

Where:

P(max) is the maximum power dissipated by the transistor in watts (W)

Uz is the zener voltage in volts (V)

Ic(max) is the maximum collector current in amperes (A)

# Formula 154.6
# Determining R2

$$R2 = \frac{0.6}{I_{R2}}$$

Where:

R2 is the resistance of R2 in ohms ($\Omega$)

$I_{R2}$ is the current through R2 in amperes (A)

0.6 is a constant used with silicon transistors - $U_{BE}$

# Formula 154.7
# $P_{R2}$ - Power Dissipated by R2:

$$P_{R2} = R_2 x I_{R2}^2$$

Where:

$P_{R2}$ is the power dissipated by R2 in watts (W)

R2 is the resistance in ohms ($\Omega$)

$I_{R2}$ is the current through R2 in amperes (A)

# Formula 154.8
# Calculating R1

R1(max)

$$R1(max) = \frac{Uin(min) - Uz - 0.6}{Iz(min) + Ic(min) + I_L(max)}$$

R1(min)

$$R1(min) = \frac{Uin(max) - Uz - 0,6}{Iz(max) - Ic(max)}$$

R1

$$R1(min) \leq R1 \leq R1(max)$$

Where:

R1(min) is the minimum value of R1 in ohms ($\Omega$)

R1(max) is the maximum value of R1 in ohms ($\Omega$)

R1 is the recommended value of R1 in ohms ($\Omega$)

Uin(min) is the minimum input voltage in volts (V)

Uin(max) is the maximum input voltage in volts (V)

Uz is the zener voltage in volts (V)

Iz(min) is the minimum zener current in amperes (A)

Iz(max) is the maximum zener current in amperes (A)

Ic(min) is the minimum collector current in amperes (A)

Ic(max) is the maximum collector current in amperes (A)

0.6 is constant - valid for silicon transistors

## Formula 154.9
## $P_{R1}$ - Dissipation of R1

$$P_{R1} = R1 x I^2$$

Where:

$P_{R1}$ is the power dissipated by R1 in watts (W)

R1 is the resistance in ohms ($\Omega$)

I is the total current in the circuit in amperes (A)

# INTEGRATED VOLTAGE REGULATORS

There are many types of integrated circuits intended for voltage regulation. The three-terminal voltage regulators are the most popular, as they are simple to use and include all the elements for protection and a high-quality voltage output.

We can find two types of three-terminal voltage regulators: fixed output or adjustable.

The next formulas are suitable when using both types: changing the original output voltage in the fixed voltage types or calculating the desired voltage in the variable types.

We also include formulas used in calculations to convert voltage regulators into current regulators (constant-current sources).

# 155. VOLTAGE REGULATORS

The basic configuration of a three-terminal voltage regulator is shown in *Figure 149*. Uin must be 2 or more volts higher than the input voltage and R1 is normally given by the manufacturer.

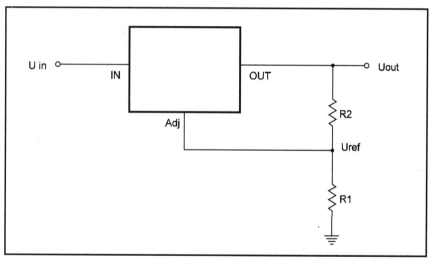

*Figure 149*

**Formula 155.1**
**Output Voltage**

$$Uout = Urefx\left(1 + \frac{R2}{R1}\right) + IadjxR2$$

Where:

>   Uout is the output voltage in volts (V)
>
>   Uref is the reference voltage in volts (V)
>
>   R1 and R2 are the resistances in ohms ($\Omega$)
>
>   Idj is the adjustment current in amperes (A)

NOTE: 1.) For the adjustable regulators Uref is normally the minimum output voltage; for instance, a 1.2 to 33V IC has a Uref of 1.2 or 1.25V (the internal zener diode).

2.) In practical cases the term IdjxR2 is very low when compared with the other terms in the formula, so it can be disconsidered without affecting practical results in a project. See next formula:

## Formula 155.2
## Simplified Formula

$$Uout = Uref\left(1 + \frac{R2}{R1}\right)$$

Where:

>   Uout is the output cvoltage in volts (V)
>
>   Uref is the reference voltage in volts (V)
>
>   R1 and R2 are the resistances in ohms ($\Omega$)

**Application Example:**

Calculate R2 when using an LM150 in a power supply for an output of 12V. R2 is 240 ohms and Uref is 1.25V.

Data:

>   Uout = 12V
>
>   Uref = 1.25V
>
>   R1 = 240 ohms
>
>   R2 = ?

Using Formula 155.2:

$$Uout = Uref\left(1 + \frac{R2}{R1}\right)$$

$$12 = 1.25x\left(1 + \frac{R2}{240}\right)$$

$$12 = 1.25x\left(\frac{240 + R2}{240}\right)$$

$$\frac{12}{1.25} = \frac{240 + R2}{240}$$

Isolating R2:

$$R2 + 240 = \frac{12x240}{1.25}$$

$$R2 = 2304 - 240 = 2064 ohms$$

# 156. CURRENT REGULATOR

The configuration shown in *Figure 150* is used to maintain constant current flowing through a load. The value of R1 is limited by the output voltage. The following formulas can be used to calculate the components and voltages in these circuits.

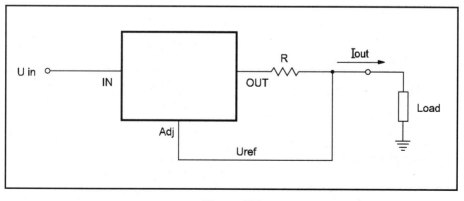

*Figure 150*

## Formula 156.1
## Resistor R

$$R = \frac{Uref}{Iout}$$

Where:

R is the resistance in ohms ($\Omega$)

Uref is the reference voltage in volts (V)

Iout is the constant current in the load in amperes (A)

**Application Example:**

Determine R for a 100 mA constant-current source using a 7805 IC (Uref = 5V) as shown in *Figure 151*.

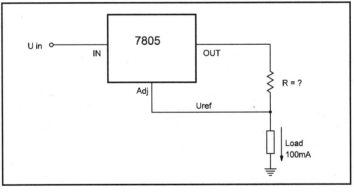

*Figure 151*

Data:

Iout = 100 mA = 0.1A

Uref = 5V

R = ?

Calculating R from Formula 156.1:

$$R = \frac{Uref}{Iout} = \frac{5}{0.1} = 50 ohms$$

# TABLE 40

## 3-Terminal Voltage Regulator ICs

| Type | Polarity | Maximum Output Current (A) | Output Voltages (V) | Reference Voltage (V) | Iadj (uA) |
|---|---|---|---|---|---|
| LM109 LM309 | Positive | 1 | 5 | 5 | - |
| LM117 LM317 | Positive | 1.5 | 1.2 to 37 | 1.25 | 50 |
| LM123A/123 LM323A/323 | Positive | 3 | 5 | 5 | - |
| LM120/320 | Negative | 1.5 | -1.2 to -47 | 1.25 | - |
| LM133/333 | Negative | 3 | -1.2 to -32 | 1.25 | 100 |
| LM137/337 | Negative | 1.5 | -1.2 to -37 | 1.25 | 65 |
| LM138A/138 LM338A/338 | Positive | 5 | 1.2 to 32 | 1.25 | 45 |
| LM140A/140 LM340A/340 | Positive | 1.5 | 5/12/15 | 5/12/15 | - |
| LM150A/150 LM350A/350 | Positive | 3 | 1.2 to 33 | 1.25 | 50 |
| LM196/396 | Positive | 10 | 1.2 to 15 | 1.25 | 50 |
| LM341 | Positive | 0.5 | 5/12/15 | 5/12/15 | - |
| 78XX series | Positive | 1 | 5/6/8/9/12/ 15/18/24 | 5/6/8/9/12/ 15/18/24 | - |
| 79XX series | Negative | 1 | 5/6/8/9/12/ 15/18/24 | 5/6/8/9/12/ 15/18/24 | - |

# Part 4

# Digital

# DIGITAL

A digital circuit follows the rules of binary arithmetic. In that numbering system, where a base 2 is used rather than the familiar 10 of the decimal system, only two digits are needed to represent any quantity. These digits are designated 1 and 0, but since a two-state circuit also follows the laws of logic, such circuits are often called *digital* circuits or *logic* circuits. The next formulas are used when working with these circuits, allowing the designer to preview what will happen with a determinated configuration or to project the desired one.

## 157. BINARY-TO-DECIMAL CONVERSION

Digital circuits use base 2. This means only two digits are used to represent a quantity. To convert a pure digital number to an equivalent decimal (base 10), use the next formula.

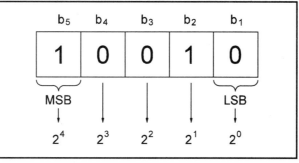

Figure 152

**Formula 157.1**
**Pure Binary-to-Decimal Conversion**

$$Dn = b1 x 2^0 + b2 x 2^1 + b3 x 2^{\cdot}$$

Where:

      Dn is the decimal number

      b1 is the least significant bit (LSB) of the binary number

      b2 to bn-1 are the intermediate bits of the binary number

      bn is the most significant digit (MSB) of the binary number

      $2^0$ to $2^n$ are powers of two (see Table 41)

**Application Example:**

Convert pure binary number 1010100 to decimal:

Applying Formula 157.1:  (LSB=0 and MSB=1)

$$Dn = 0x2^0 + 0x2^1 + 1x2^2 + 0x2^3 + 1x2^4 + 0x2^5 + 1x2^6$$
$$Dn = 0x1 + 0x2 + 1x4 + 0x8 + 1x16 + 0x32 + 1x64$$
$$Dn = 4 + 16 + 64$$
$$Dn = 80$$

## TABLE 41
## Powers of Two

| Binary | Decimal | Binary | Decimal |
|--------|---------|--------|---------|
| $2^0$ | 1 | $2^{17}$ | 131 072 |
| $2^1$ | 2 | $2^{18}$ | 262 144 |
| $2^2$ | 4 | $2^{19}$ | 524 288 |
| $2^3$ | 8 | $2^{20}$ | 1 048 576 |
| $2^4$ | 16 | $2^{21}$ | 2 097 152 |
| $2^5$ | 32 | $2^{22}$ | 4 194 304 |
| $2^6$ | 64 | $2^{23}$ | 8 388 608 |
| $2^7$ | 128 | $2^{24}$ | 16 777 216 |
| $2^8$ | 256 | $2^{25}$ | 33 554 432 |
| $2^9$ | 512 | $2^{26}$ | 67 108 864 |
| $2^{10}$ | 1 024 | $2^{27}$ | 134 217 728 |
| $2^{11}$ | 2 048 | $2^{28}$ | 268 435 456 |
| $2^{12}$ | 4 096 | $2^{29}$ | 536 870 912 |
| $2^{13}$ | 8 192 | $2^{30}$ | 1 073 741 824 |
| $2^{14}$ | 16 384 | $2^{31}$ | 2 147 483 648 |
| $2^{15}$ | 32 768 | $2^{32}$ | 4 294 967 296 |
| $2^{16}$ | 65 536 | | |

# 158. BYTE-TO-DECIMAL CONVERSION

The byte is an 8-bit binary number. The next formula can be used to convert a byte in decimal:

**Formula 158.1**

**Byte-to-Decimal**

$$Dn = b1x2^0 + b2x^1 + b3x2^2 + b4x2^3 + b5x2^4 + b6x2^5 + b7x2^6 + b8x2^7$$

Where:

Dn is the decimal number

b1 to b8 are the bits of the byte

b1 is the MSB (most significant bit)

b2 is the LSB (least significant bit)

**TABLE 42**

**Decimal Integers to Pure Binaries**

| Decimal | Binary | Decimal | Binary | Decimal | Binary |
|---------|--------|---------|--------|---------|--------|
| 00 | 00000000 | 13 | 00001101 | 26 | 00011010 |
| 01 | 00000001 | 14 | 00001110 | 27 | 00011011 |
| 02 | 00000010 | 15 | 00001111 | 28 | 00011100 |
| 03 | 00000011 | 16 | 00010000 | 29 | 00011101 |
| 04 | 00000100 | 17 | 00010001 | 30 | 00011110 |
| 05 | 00000101 | 18 | 00010010 | 31 | 00011111 |
| 06 | 00000110 | 19 | 00010011 | 32 | 00100000 |
| 07 | 00000111 | 20 | 00010100 | 33 | 00100001 |
| 08 | 00001000 | 21 | 00010101 | 34 | 00100010 |
| 09 | 00001001 | 22 | 00010110 | 35 | 00100011 |
| 10 | 00001010 | 23 | 00010111 | 36 | 00100100 |
| 11 | 00001011 | 24 | 00011000 | 37 | 00100101 |
| 12 | 00001100 | 25 | 00011001 | 38 | 00100110 |

(continued next page)

| Decimal | Binary | Decimal | Binary | Decimal | Binary |
|---------|----------|---------|----------|---------|----------|
| 39 | 00100111 | 50 | 00110010 | 61 | 00111101 |
| 40 | 00101000 | 51 | 00110011 | 62 | 00111110 |
| 41 | 00101001 | 52 | 00110100 | 63 | 00111111 |
| 42 | 00101010 | 53 | 00110101 | 64 | 01000000 |
| 43 | 00101011 | 54 | 00110110 | 65 | 01000001 |
| 44 | 00101100 | 55 | 00110111 | 66 | 01000010 |
| 45 | 00101101 | 56 | 00111000 | 67 | 01000011 |
| 46 | 00101110 | 57 | 00111001 | 68 | 01000100 |
| 47 | 00100111 | 58 | 00111010 | 69 | 01000101 |
| 48 | 00110000 | 59 | 00111011 | 70 | 01000110 |
| 49 | 00110001 | 60 | 00111100 | | |

# 159. BCD TO DECIMAL

Binary-coded decimal is a form of binary representation used in digital electronics where groups of 4 bits represent a decimal digit, as shown in Figure 153. Conversion to decimal is made as follows:

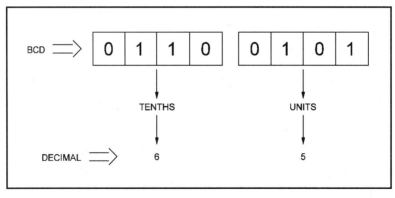

*Figure 153*

# Formula 159.1
# BCD-to-Decimal Conversion

$$Dd = b1x2^0 + b2x2^1 + b3x2^2 + b4x2^3$$
$$and$$
$$Dd < 10$$

Where:

Dd is the decimal digit

b1 to b4 are the BCD digits or bits

b1 is the MSB (most significant bit)

b4 is the LSB (least significant bit)

**Application Example:**

Convert to decimal the BCD 1001 0100.

Solving:

Calculating the units decimal digit:

$$Ddu = 0x2^0 + 0x2^1 + 1x2^2 + 0x2^3$$
$$Ddu = 0 + 0 + 8 + 0$$
$$Ddu = 8$$

Calculating the tenths decimal digit:

$$Ddt = 1x2^0 + 0x2^1 + 0x2^2 + 1x2^3$$
$$Ddt = 1 + 8$$
$$Ddt = 9$$

The decimal number is 98.

**TABLE 43**
**Negative Powers of Two**

| Negative Power of Two | Decimal |
|---|---|
| $2^0$ | 1 |
| $2^{-1}$ | 0.5 |
| $2^{-2}$ | 0.25 |
| $2^{-3}$ | 0.125 |
| $2^{-4}$ | 0.062 5 |
| $2^{-5}$ | 0.031 25 |
| $2^{-6}$ | 0.015 625 |
| $2^{-7}$ | 0.007 812 5 |
| $2^{-8}$ | 0.003 906 25 |
| $2^{-9}$ | 0.001 953 125 |
| $2^{-10}$ | 0.000 976 562 5 |
| $2^{-11}$ | 0.000 488 281 25 |
| $2^{-12}$ | 0.000 244 140 625 |
| $2^{-13}$ | 0.000 122 070 312 5 |
| $2^{-14}$ | 0.000 061 035 156 25 |
| $2^{-15}$ | 0.000 030 517 578 125 |
| $2^{-16}$ | 0.000 015 258 789 062 5 |

# 160. HEXADECIMAL-TO-DECIMAL CONVERSION

In this numbering system, digits 0 through 9 are used and also letters A to F. As in the case of binary and decimal numbers, the value of a hexadecimal number depends on its horizontal position. Conversion to decimal is made using the next formula. Table 44 gives the values of each digit in the hexadecimal numbering system.

# Formula 160.1
# Hexadecimal-to-Decimal Conversion

$$Dn = h1x16^2 + h2x16^1 + h3x16^2 + h4x16^4$$

Where:

Dn is the decimal number

h1 to h4 are the hexadecimal digits (*)

h1 is the LSB hexadecimal digit

h4 is the MSB hexadecimal digit

(* ) See Table 44 to know the equivalent decimal digits to hexadecimal letters.

**Application Example:**

Convert to decimal the hexadecimal F5A2.

Data:

h1 = 2

h2 = A (10)

h3 = 5

h4 = F (15)

Applying the formula and consulting Table 45 to powers of 16:

$$Dn = 2x16^0 + 10x16^1 + 5x16^2 + 15x16^3$$
$$Dn = 2x1 + 10x16 + 5x256 + 15x4096$$
$$Dn = 2 + 160 + 1280 + 61440$$
$$Dn = 62882$$

**TABLE 44**

**Hexadecimal Digits and Decimal Correspondents**

| Hexadecimal | Decimal | Hexadecimal | Decimal |
|---|---|---|---|
| 0 | 0 | 8 | 8 |
| 1 | 1 | 9 | 9 |
| 2 | 2 | A | 10 |
| 3 | 3 | B | 11 |
| 4 | 4 | C | 12 |
| 5 | 5 | D | 13 |
| 6 | 6 | E | 14 |
| 7 | 7 | F | 15 |

**TABLE 45**

**Powers of 16**

| Power of 16 | Decimal |
|---|---|
| $16^0$ | 1 |
| $16^1$ | 16 |
| $16^2$ | 256 |
| $16^3$ | 4096 |
| $16^4$ | 65 536 |
| $16^5$ | 1 048 576 |
| $16^6$ | 16 777 216 |

# 161. DECIMAL-TO-BINARY CONVERSION

There isn't a formula to make this conversion. To convert a decimal number to pure binary we have to use an algorithm consisting in successive divisions of the decimal number by the binary base (2). How to use this algorithm is shown in the next lines.

## Algorithm 161.1
## Converting Decimal-to-Binary

$$b1.......bn = \left[\frac{dn}{2}\right]\bar{R}$$

"The binary number is found by writing in the inverse order the rest of the successive division of the decimal number by two, beginning by the result of the last division."

Where:

b1 to b1 is the binary number

dn is the decimal number

$\bar{R}$ is the result of the divisions in inverse order

**Application Example:**

Convert the decimal number 26 to binary:

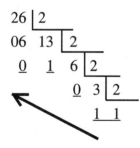

Writing in the inverse order: 11010.

# LOGIC FUNCTIONS

The logic functions are blocks used in digital circuits to perform a series of "yes-no" decisions based on the presence or absence of signals on various inputs. The input signals are other "yes and no" signals, and depending on how the blocks are interconnected it is possible to build a simple timer through a complex computer.

The following information, including tables, functions, or relationships between the blocks, will be useful for the designer working on projects and calculations, or even those who know how a digital block operates in an application.

# 162. AND GATE

The output of an AND gate is in the high logic level only if either input (A or B) is in the high logic level. *Figure 154* shows the symbol of an AND gate and the equivalent circuit.

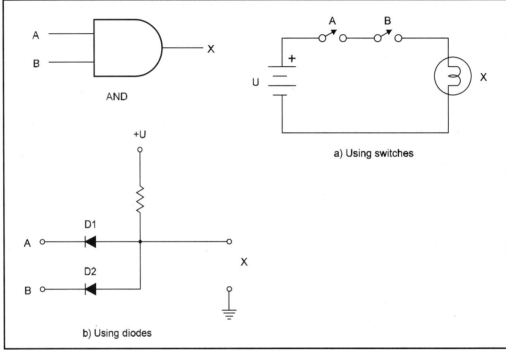

*Figure 154*

## Equation 162.1
## Boolean Equation — 2-Input AND Gate

$$X = A.B$$

Where:

X is the output logic level

A, B are the input logic levels

# TABLE 46
# Truth Table - 2-Input AND Function

| A | B | X |
|---|---|---|
| 0 | 0 | 0 |
| 0 | 1 | 0 |
| 1 | 0 | 0 |
| 1 | 1 | 1 |

# Equation 161.2
# Boolean Equation - 3-Input AND Gate

$$X = ABC$$

Where:

X is the output logic level

A, B and C are input logic levels

# TABLE 47
# Truth Table - 3-Input NAND Gate

| A | B | C | X |
|---|---|---|---|
| 0 | 0 | 0 | 0 |
| 0 | 0 | 1 | 0 |
| 0 | 1 | 0 | 0 |
| 0 | 1 | 1 | 0 |
| 1 | 0 | 0 | 0 |
| 1 | 0 | 1 | 0 |
| 1 | 1 | 0 | 0 |
| 1 | 1 | 1 | 1 |

# 163. NAND GATE

The output is high only if the inputs (A and B) are not in the high level. The symbol and the equivalent circuit is shown in *Figure 155*.

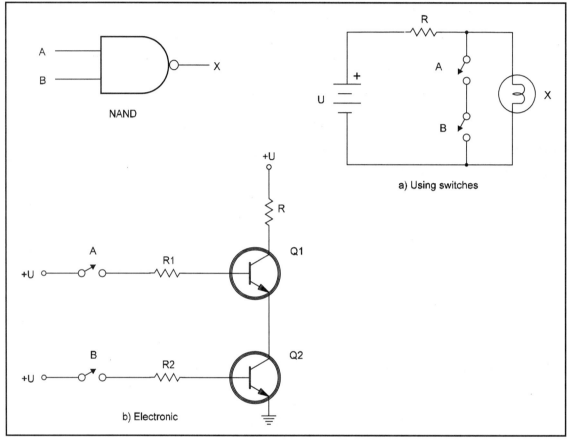

*Figure 155*

## Equation 163.1

## Boolean Equation — 2-Input NAND Gate

$$X = \overline{AB} = \overline{A} + \overline{B}$$

Where:

      X is the output logic level

      A and B are the input logic levels

# TABLE 48
# Truth Table — 2-Input NAND Gate

| A | B | X |
|---|---|---|
| 0 | 0 | 1 |
| 0 | 1 | 1 |
| 1 | 0 | 1 |
| 1 | 1 | 0 |

# Equation 163.2
# Boolean Equation - 3-Input NAND Gate

$$X = \overline{ABC} = \overline{A} + \overline{B} + \overline{C}$$

Where:

X is the output logic level

A, B and C are the input logic levels

# TABLE 49
# Truth Table — 3-Input NAND Gate

| A | B | C | X |
|---|---|---|---|
| 0 | 0 | 0 | 1 |
| 0 | 0 | 1 | 1 |
| 0 | 1 | 0 | 1 |
| 0 | 1 | 1 | 1 |
| 1 | 0 | 0 | 1 |
| 1 | 0 | 1 | 1 |
| 1 | 1 | 0 | 1 |
| 1 | 1 | 1 | 0 |

# 164. OR GATE

The output of an OR gate is high if either input, A or B, or both, are in the high logic level. *Figure 156* shows the symbol and the equivalent circuit.

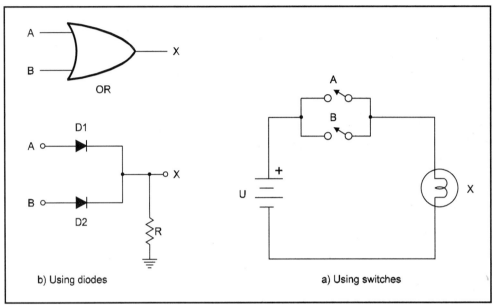

*Figure 156*

### Equation 164.1
### Boolean Equation — 2-Input OR Gate

$$X = A + B$$

Where:

X is the output logic level

A and B are the input logic levels

# TABLE 50
## Truth Table — 2-Input OR Gate

| A | B | X |
|---|---|---|
| 0 | 0 | 0 |
| 0 | 1 | 1 |
| 1 | 0 | 1 |
| 1 | 1 | 1 |

# Equation 164.2
## Boolean Equation - 3-Input OR Gate

$$X = A + B + C$$

Where:

X is the output logic level

A, B and C are the input logic levels

# TABLE 51
## Truth Table — 3-Input OR Gate

| A | B | C | X |
|---|---|---|---|
| 0 | 0 | 0 | 0 |
| 0 | 0 | 1 | 1 |
| 0 | 1 | 0 | 1 |
| 0 | 1 | 1 | 1 |
| 1 | 0 | 0 | 1 |
| 1 | 0 | 1 | 1 |
| 1 | 1 | 0 | 1 |
| 1 | 1 | 1 | 1 |

# 165. NOR GATE

The output of a NOR gate is at the high logic level if neither A nor B are in the high logic level. *Figure 157* shows the symbol and the equivalent circuit.

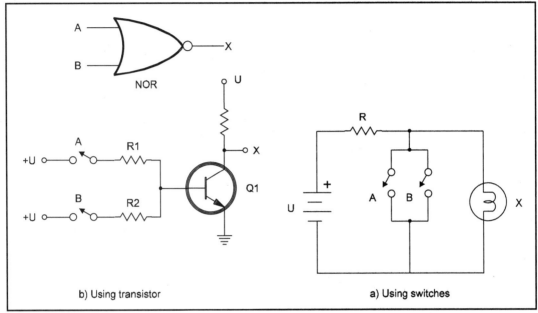

b) Using transistor          a) Using switches

*Figure 157*

## Equation 165.1
## Boolean Equation — 2-Input NOR Gate

$$X = \overline{A + B}$$

Where:

X is the output logic level

A and B are the input logic level

## TABLE 52
## Truth Table — 2-Input NOR Gate

| A | B | X |
|---|---|---|
| 0 | 0 | 1 |
| 0 | 1 | 0 |
| 1 | 0 | 0 |
| 1 | 1 | 0 |

## Equation 165.2
## Boolean Equation — 3-Input NOR Gate

$$X = \overline{A + B + C}$$

Where:

X is the output logic level

A, B and C are the input logic level

## TABLE 53
## Truth Table — 3-Input NOR Gate

| A | B | C | X |
|---|---|---|---|
| 0 | 0 | 0 | 1 |
| 0 | 0 | 1 | 0 |
| 0 | 1 | 0 | 0 |
| 0 | 1 | 1 | 0 |
| 1 | 0 | 0 | 0 |
| 1 | 0 | 1 | 0 |
| 1 | 1 | 0 | 0 |
| 1 | 1 | 1 | 0 |

# 166. EXCLUSIVE-OR

In the output of an exclusive-OR, the logic level is high if either input A or B is in the high logic level, but not both. The symbol and the equivalent circuit of a exclusive-OR gate are shown in *Figure 158*.

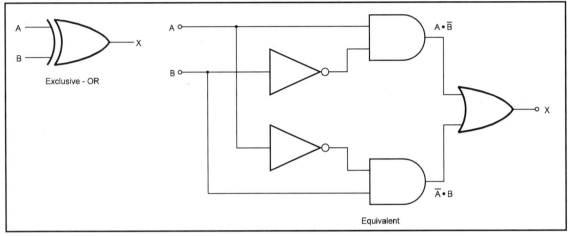

*Figure 158*

### Equation 166.1
### Boolean Equation — Exlusive-OR Gate

$$X = A\overline{B} + \overline{A}B$$

Where:

        X is the output logic level

        A and B are the input logic level

### TABLE 54
### Truth Table — Exclusive-OR Gate

| A | B | X |
|---|---|---|
| 0 | 0 | 0 |
| 0 | 1 | 1 |
| 1 | 0 | 1 |
| 1 | 1 | 0 |

# 167. INVERTER

The logic level at the output of an inverter is the opposite of the input logic level. The symbol and the equivalent circuit is shown in *Figure 159*.

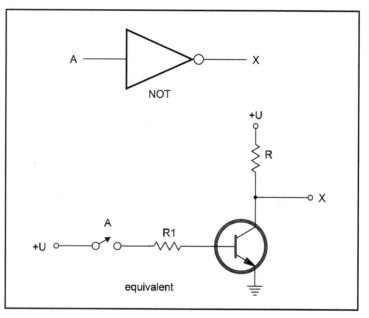

*Figure 159*

## Equation 167.1
## Inverter

$$X = \overline{A}$$

Where:

X is the output logic level

A is the input logic level

# 168. BINARY ADDITION

The next general rules are valid when making binary additions:

**Rule 168.1**
**Binary Addition**

$$0 + 0 = 0$$
$$0 + 1 = 1$$
$$1 + 0 = 1$$
$$1 + 1 = 1 \text{ and 1 to carry}$$

# 169. BINARY SUBTRACTION

The next general rules are valid when making binary subtractions:

**Rule 169.1**
**Binary Subtraction**

$$0 - 0 = 0$$
$$0 - 1 = 1 \text{ and 1 to borrow}$$
$$1 - 0 = 0$$
$$1 - 1 = 0$$

# 170. BINARY MULTIPLICATION

The next rules are valid when making binary multiplications:

**Rule 170.1**
**Binary Multiplication**

$$0 \times 0 = 0$$
$$0 \times 1 = 0$$
$$1 \times 0 = 0$$
$$1 \times 1 = 1$$

# 171. BINARY DIVISION

The next rules are valid when making binary divisions:

**Rule 171.1**
**Binary Division**

$$0/0 = ?$$
$$0/1 = 0$$
$$1/0 = ?$$
$$1/1 = 1$$

# THE POSTULATES OF BOOLEAN ALGEBRA

# 172. LAWS OF TAUTOLOGY

Repetition by addition or multiplication does not alter the truth value of an element. *Figure 160* shows the circuit diagram correspondent to this law.

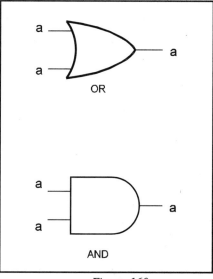

*Figure 160*

**Law 172.1**
**Tautology**

$$a + a = a$$

$$a \times a = a$$

## Mathematic Notation (Set Theory)

$$a \cup a = a$$
$$a \cap a = a$$

## Logic Notation

$$a \vee a = a$$
$$a \wedge a = a$$

# 173. LAWS OF COMMUTATION

Conjunction and disjunction are not affected by sequential change. *Figure 161* shows the circuit diagrams correspondent to this law.

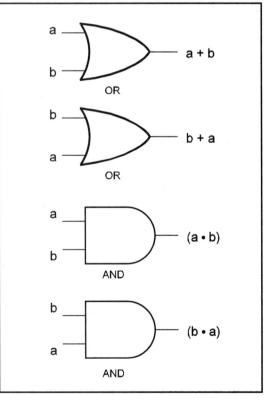

*Figure 161*

## Law 173.1
## Commutation

$$a + b = b + a$$
$$axb = bxa$$

## Mathematic Notation

$$a \cup b = b \cup a$$
$$a \cap b = b \cap a$$

## Logic Notation

$$a \vee b = b \vee a$$
$$a \wedge b = b \wedge a$$

# 174. LAWS OF ASSOCIATION

Grouping does not affect disjunction or conjunction  *Figure 162* shows equivalent diagrams.

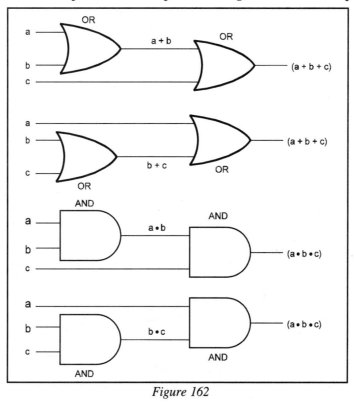

*Figure 162*

## Laws 174.1
## Association Laws

$$a + (b + c) = (a + b) + c$$
$$ax(bxc) = (axb)xc$$

### Mathematic Notation

$$a \cup (b \cup c) = (a \cup b) \cup c$$
$$a \cap (b \cap c) = (a \cap b) \cap c$$

### Logic Notation

$$a \vee (b \vee c) = (a \vee b) \vee c$$
$$a \wedge (b \wedge c) = (a \wedge b) \wedge c$$

# 175. LAWS OF DISTRIBUTION

An element is added to a product by adding the element to each member of the product and a sum is multiplied by an element by multiplying every member of the sum by the element. The circuit diagram correspondent to these laws is shown in *Figure 163*.

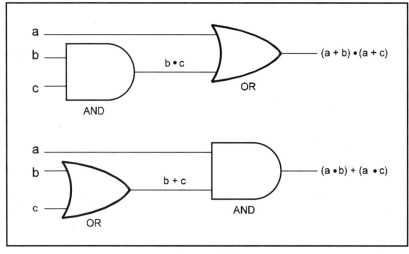

*Figure 163*

## Laws 175.1
## Distribution Laws

$$a + (bxc) = (a + b)x(a + c)$$
$$ax(b + c) = (axb) + (axc)$$

## Mathematic Notation

$$a \cup (b \cap c) = (a \cup b) \cap (a \cup c)$$
$$a \cap (b \cup c) = (a \cap b) \cup (a \cap c)$$

## Logic Notation

$$a \vee (b \wedge c) = (a \vee b) \wedge (a \vee c)$$
$$a \wedge (b \vee c) = (a \wedge b) \vee (a \wedge c)$$

# 176. LAWS OF ABSORPTION

The disjunction of a product by one of its members is equivalent to this member. The conjunction of a sum by one of its members is equivalent to this member. The equivalent logic circuit is shown in *Figure 164*.

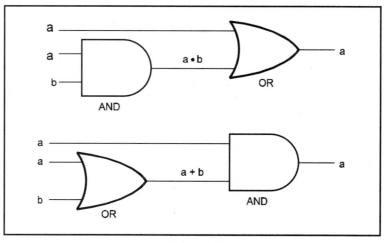

*Figure 164*

## Laws 176.1
## Absorption Laws

$$a + (axb) = a$$
$$ax(a + b) = a$$

### Mathematic Notation

$$a \cup (a \cap b) = a$$
$$a \cap (a \cup b) = a$$

### Logic Notation

$$a \vee (a \wedge b) = a$$
$$a \wedge (a \vee b) = a$$

# 177. LAWS OF UNIVERSE CLASS

The sum consisting of an element and the universe class is equivalent to the universe class. The product consisting of an element and the universe class is equivalent to the element. The equivalent circuits are shown in *Figure 165*.

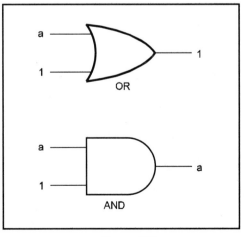

*Figure 165*

## Laws 177.1
## Universe Class

$$a + 1 = 1$$
$$ax1 = a$$

## Mathematic Notation

$$a \cup 1 = 1$$
$$a \cap 1 = a$$

## Logic Notation

$$a \vee 1 = 1$$
$$a \wedge 1 = a$$

# 178. LAWS OF NULL CLASS

The sum consisting of an element and the null class is equivalent to the element. The product consisting of an element and the null class is equivalent to the null class. The equivalent logic circuits are shown in *Figure 166*.

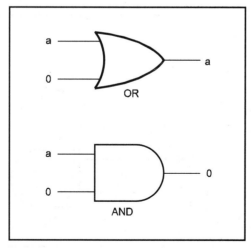

*Figure 166*

## Laws 178.1
## Null Class

$$a + 0 = a$$
$$ax0 = 0$$

## Mathematic Notation

$$a \cup 0 = a$$
$$a \cap 0 = 0$$

## Logic Notation

$$a \vee 0 = a$$
$$a \wedge 0 = 0$$

# 179. LAWS OF COMPLEMENTATION

The sum consisting of an element and its complement is equivalent to the universe class. The product consisting of an element and its complement is equivalent to the null class. *Figure 167* shows the logic diagrams correspondent to these laws.

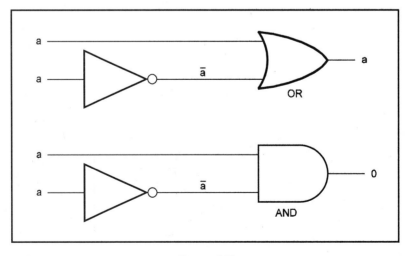

*Figure 167*

## Laws 179.1
## Complementation

$$a + \overline{a} = 1$$
$$ax\overline{a} = 0$$

## Mathematic Notation

$$a \cup \overline{a} = 1$$
$$a \cap \overline{a} = 0$$

## Logic Notation

$$a \vee \overline{a} = 1$$
$$a \wedge \overline{a} = 0$$

# 180. LAWS OF CONTRAPOSITION

If an element **a** is equivalent to the complement of an element **b** it is implied that the element **b** is equivalent to the complement of the element **a**. *Figure 168* shows the equivalent diagram.

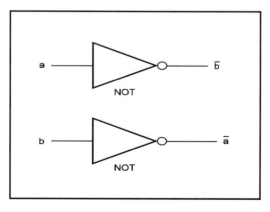

*Figure 168*

**Law 180.1**

**Contraposition**

$$a = \bar{b} \Rightarrow b = \bar{a}$$

**Mathematic Notation**

$$a = b' \geq b = a'$$

**Logic Notation**

$$a \equiv\approx b . \geq . b \equiv\approx a$$

# 181. LAW OF DOUBLE NEGATION

The complement of the negation of an element is equivalent to the element. *Figure 169* shows the equivalent circuit diagram.

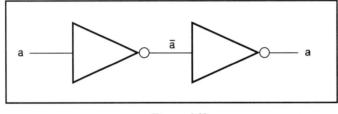

*Figure 169*

**Law 181.1**

**Double Negation**

$$a = \bar{\bar{a}}{}'$$

**Mathematic Notation**

$$a = a'\, C$$

**Logic Notation**

$$a =\sim a'$$

# 182. LAWS OF EXPANSION

The disjunction of a product composed of the elements **a** and **b** and a product composed of the element **a** and the complement of the element **b** is equivalent to the element **a**. The conjunction of a sum composed of the elements **a** and **b** and a sum composed of the element **a** and the complement of the element **b** is equivalent to the element **a**. *Figure 170* shows the circuit diagrams describing these laws.

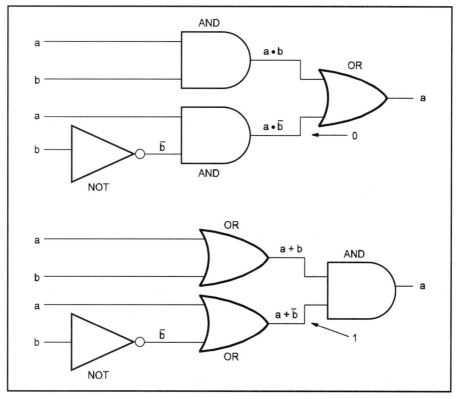

*Figure 170*

**Law 182.1**

**Laws of Expansion**

$$(axb) + (ax\overline{b}) = a$$

$$(a+b)x(a+\overline{b}) = a$$

## Mathematic Notation

$$(a \cap b) \cup (a \cap b') = a$$
$$(a \cup b) \cap (a \cup b') = a$$

## Logic Notation

$$(a \wedge b) \vee (a \wedge \sim b) = a$$
$$(a \vee b) \wedge (a \vee \sim b) = a$$

# 183. LAWS OF DUALITY

The complement of a sum composed of the elements **a** and **b** is equivalent to the conjunction of the complement of the element **a** and the complement of the element **b**. The complement of a product composed of the elements **a** and **b** is equivalent to the disjunction of the complement of element **a** and the complement of element **a**. *Figure 171* shows the equivalent circuit.

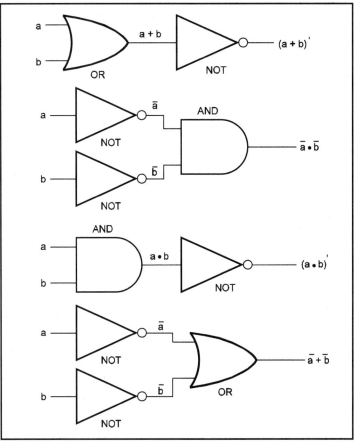

*Figure 171*

## Law 183.1
## Law of Duality

$$(a + b)' = \overline{a}x\overline{b}$$
$$(axb)' = \overline{a} + \overline{b}$$

## Mathematic Notation

$$(a \cup b)' = a' \cap b'$$
$$(a \cap b)' = a' \cup b'$$

## Logic Notation

$$\sim (a \vee b) = \sim a \wedge \sim b$$
$$\sim (a \wedge b) = \sim a \vee \sim b$$

# BOOLEAN RELATIONSHIPS

# 184. IDEMPOINT

## Relationship 184.1
## Addition

$$a + 0 = a$$
$$a + 1 = 1$$
$$a + a = a$$

Where: $0 \equiv a$

## Relationship 184.2
## Multiplication

$$0xa = 0$$
$$1xa = a$$
$$axa = a$$

Where: $0 \equiv a$

# 185. COMMUTATIVE

**Relationship 185.1**
**Addition**

$$(a+b) = (b+a)$$

**Relationship 185.2**
**Multiplication**

$$axb = bxa$$

# 186. ASSOCIATIVE

**Relationship 186.1**
**Addition**

$$(a+b) + c = a + (b+c)$$

**Relationship 186.2**
**Multiplication**

$$(axb)xc = ax(bxc)$$

# 187. DISTRIBUTIVE

**Relationship 187.1**
**Distributive**

$$a + (bxc) = ax(b+c)$$
$$a + bxc = (a+b)x(a+c)$$

# 188. ABSORPTION

**Relationship 176.1**

$$ax(a+b) = a + axb \equiv a$$

# 189. DEMORGAN THEOREM

## Relationship 189.1
## DeMorgan

$$\overline{\overline{a}} = a$$

$$Theorem-1:$$

$$\overline{(axb)} = \overline{a} + \overline{b}$$

$$\overline{\overline{(axb)}} = a + b$$

$$Theorem-2:$$

$$\overline{a+b} = \overline{a}x\overline{b}$$

$$\overline{\overline{a}+\overline{b}} = axb$$

## TABLE 55
## Some TTL AND/NAND Gates

| Type | Function |
|------|----------|
| 7400 | Quad, 2-input NAND |
| 7401 | Quad, 2-input, NAND,Open Collector |
| 7408 | Quad -2-input, AND |
| 7409 | Quad - 2-input, AND, Open Collecor |
| 7410 | Triple, 3-input, NAND |
| 7411 | Triple, 3-input, AND |
| 7412 | Triple, 3-input, NAND, Open Collector, $V_{OH}$=5.5 V |
| 7403 | Quad, 2-input, NAND, Open Collector, $V_{OH}$=5.5 V |
| 7420 | Dual, 4-input, NAND |
| 7426 | Quad, 2-input, NAND. Open Collector |
| 7437 | Quad, 2-input, NAND, $I_{OL}$=48 mA |
| 7438 | Quad, 2-input, NAND, Open Collector |

## TABLE 56
## Common TTL OR/NOR Gates

| Type | Function |
|------|----------|
| 7402 | Quad, 2-input, NOR |
| 7425 | Dual, 4-input, NOR |
| 7427 | Triple, 3-input, NOR |
| 7432 | Quad, 2-input, OR |

## TABLE 57
## TTL Subfamily Fan-Out Rules

| Family/Subfamily | Will Drive....Iinputs |
|------------------|------------------------|
| Regular TTL | 10 Regular TTL |
| | 40 Low-power TTL |
| | 6 High-power TTL |
| | 6 Schottky TTL |
| | 20 Low-power Schottky TTL |
| Low-Power TTL | 2 Regular TTL |
| | 10 Low-power TTL |
| | 1 High-power TTL |
| | 1 Schottky TTL |
| | 5 Low-power Schottky TTL |
| High-Power TTL | 12 Regular TTL |
| | 40 Low-power TTL |
| | 10 High-power TTL |
| | 10 Schottky TTL |
| | 40 Low-power Schottky TTL |

(continued next page)

| Family/Subfamily | Will Drive....Iinputs |
|---|---|
| Schottky TTL | 12 Regular TTL |
| | 40 Low-power TTL |
| | 10 High-power TTL |
| | 10 Schottky TTL |
| | 40 Low-power Schottky TTL |
| Low-power Schottky | 5 Regular TTL |
| | 20 Low-power TTL |
| | 4 High-power TTL |
| | 4 Schottky TTL |
| | 10 Low-power Schottky TTL |

# Part 5

# Miscellaneous

# MISCELANEOUS

# 190. INSTANTANEOUS VALUES OF SINE WAVES

The next formula is used to calculate the instanteneous value of a sine wave voltage or current from the peak value and the phase angle. The table below is also used in the calculations to give the instantaneous relative value of a sine wave as a function of the peak value and the phase angle.

## Formula 190.1
## Instantaneous Value

$$V = Vpx\frac{Tv}{100}$$

Where:

V is the instantaneous value (voltage, for instance)

Vp is the peak value (voltage for instance)

Tv is the value found in the table for the desired phase angle

## TABLE 58
## Instantaneous Values of Sine Wave (Current or Voltage)

| Phase Angle (Degrees) | | | | Percent (Tv) |
|---|---|---|---|---|
| 0 | 180 | 180 | 360 | 0.000000 |
| 1 | 179 | 181 | 359 | 1.745241 |
| 2 | 178 | 182 | 358 | 3.489950 |
| 3 | 177 | 183 | 357 | 5.233596 |
| 4 | 176 | 184 | 356 | 6.975647 |
| 5 | 175 | 185 | 355 | 8.715574 |
| 6 | 174 | 186 | 354 | 10.45285 |
| 7 | 173 | 187 | 353 | 12.18693 |
| (continued next page) | | | | |

| Phase Angle (Degrees) | | | | Percent (Tv) |
|---|---|---|---|---|
| 8 | 172 | 188 | 352 | 13.92731 |
| 9 | 171 | 189 | 351 | 15.64345 |
| 10 | 170 | 190 | 350 | 17.36482 |
| 11 | 169 | 191 | 349 | 19.08090 |
| 12 | 168 | 192 | 348 | 20.79117 |
| 13 | 167 | 193 | 347 | 22.49511 |
| 14 | 166 | 194 | 346 | 24.19219 |
| 15 | 165 | 195 | 345 | 25.88190 |
| 16 | 164 | 196 | 344 | 27.56374 |
| 17 | 163 | 197 | 343 | 29.23717 |
| 18 | 162 | 198 | 342 | 30.90170 |
| 19 | 161 | 199 | 341 | 32.55682 |
| 20 | 160 | 200 | 340 | 34.20201 |
| 21 | 159 | 201 | 339 | 35.83680 |
| 22 | 158 | 202 | 338 | 37.46066 |
| 23 | 157 | 203 | 337 | 39.07311 |
| 24 | 156 | 204 | 336 | 40.67366 |
| 25 | 155 | 205 | 335 | 42.26183 |
| 26 | 154 | 206 | 334 | 43.83711 |
| 27 | 153 | 207 | 333 | 45.39905 |
| 28 | 152 | 208 | 332 | 46.94715 |
| 29 | 151 | 209 | 331 | 48.48096 |
| 30 | 150 | 210 | 330 | 50.00000 |
| 31 | 149 | 211 | 329 | 51.50381 |
| 32 | 148 | 212 | 328 | 52.99192 |
| 33 | 147 | 213 | 327 | 54.46391 |
| 34 | 146 | 214 | 326 | 55.91929 |
| 35 | 145 | 215 | 325 | 57.35764 |

(continued next page)

| Phase Angle (Degrees) | | | | Percent (Tv) |
|---|---|---|---|---|
| 36 | 144 | 216 | 324 | 58.77853 |
| 37 | 143 | 217 | 323 | 60.18150 |
| 38 | 142 | 218 | 322 | 61.56615 |
| 39 | 141 | 219 | 321 | 62.93204 |
| 40 | 140 | 220 | 320 | 64.27876 |
| 41 | 139 | 221 | 319 | 65.60590 |
| 42 | 138 | 222 | 318 | 66.91306 |
| 43 | 137 | 223 | 317 | 68.19984 |
| 44 | 136 | 224 | 316 | 69.46584 |
| 45 | 135 | 225 | 315 | 70.71068 |
| 46 | 134 | 226 | 314 | 71.93398 |
| 47 | 133 | 227 | 313 | 73.13537 |
| 48 | 132 | 228 | 312 | 74.31448 |
| 49 | 131 | 229 | 311 | 75.47095 |
| 50 | 130 | 230 | 310 | 76.60445 |
| 51 | 129 | 231 | 309 | 77.71460 |
| 52 | 128 | 232 | 308 | 78.80107 |
| 53 | 127 | 233 | 307 | 79.86355 |
| 54 | 126 | 234 | 306 | 80.90170 |
| 55 | 125 | 235 | 305 | 81.91521 |
| 56 | 124 | 236 | 304 | 82.90376 |
| 57 | 123 | 237 | 303 | 83.86706 |
| 58 | 122 | 238 | 302 | 84.80481 |
| 59 | 121 | 239 | 301 | 85.71673 |
| 60 | 120 | 240 | 300 | 86.60254 |
| 61 | 119 | 241 | 299 | 87.46198 |
| 62 | 118 | 242 | 298 | 88.29475 |
| 63 | 117 | 243 | 297 | 89.10065 |

(continued next page)

| Phase Angle (Degrees) | | | | Percent (Tv) |
|---|---|---|---|---|
| 64 | 116 | 244 | 296 | 89.87940 |
| 65 | 115 | 245 | 295 | 90.63078 |
| 66 | 114 | 246 | 294 | 91.35455 |
| 67 | 113 | 247 | 293 | 92.05048 |
| 68 | 112 | 248 | 292 | 92.71838 |
| 69 | 111 | 249 | 291 | 93.35804 |
| 70 | 110 | 250 | 290 | 93.96926 |
| 71 | 109 | 251 | 289 | 94.55186 |
| 72 | 108 | 252 | 188 | 95.10565 |
| 73 | 107 | 253 | 287 | 95.63048 |
| 74 | 106 | 254 | 286 | 96.12617 |
| 75 | 105 | 255 | 285 | 96.59258 |
| 76 | 104 | 256 | 284 | 97.02957 |
| 77 | 103 | 257 | 283 | 97.43700 |
| 78 | 102 | 258 | 282 | 97.81476 |
| 79 | 101 | 259 | 281 | 98.16272 |
| 80 | 100 | 160 | 280 | 98.48077 |
| 81 | 99 | 261 | 279 | 98.76884 |
| 82 | 98 | 262 | 278 | 99.02681 |
| 83 | 97 | 263 | 277 | 99.25462 |
| 84 | 96 | 264 | 276 | 99.45219 |
| 85 | 95 | 265 | 275 | 99.61497 |
| 86 | 94 | 266 | 274 | 99.75640 |
| 87 | 93 | 267 | 273 | 99.86295 |
| 88 | 92 | 268 | 272 | 99.93908 |
| 89 | 91 | 269 | 271 | 99.98477 |
| 90 | 90 | 270 | 270 | 100.00000 |

**Application Example:**

Determine the instantaneous value of a 300V (peak) voltage at the phase angle of 32 degrees.

Data:

Vpeak = 300 V

$\alpha$ = 32°

Using the Formula:

V = Vpeak x Tv/100

Where:

V is the instantaneous voltage in volts (V)

Vpeak is the peak voltage in volts (V)

Tv is the value found in the table

$$V = 300x\frac{52.99}{100} = 158.97 \ V$$

# 191. INCHES TO MILLIMETERS

One inch is equal to 25.40 millimeters. The next formulas are used to convert millimeters to inches and inches to millimeters.

## Formula 191.1
## Millimeters to Inches

$$I = 0.038907xM$$

or

$$I = \frac{M}{25.40}$$

Where:

M is the length in millimeters (mm)

I is the length in inches

## Formula 191.2
## Inches to Millimeters

$$M = 25.40xI$$

or

$$M = \frac{I}{0.03907}$$

Where:

I is the length in inches

M is the length in millimeters

**Application Example:**

Convert 3.2 inches to millimeters:

Data:

I = 3.2

M = ?

Using Formula 191.2:

M = 25.40 x 3.2 = 81.82

# TABLE 59
## Decimal Inches to Millimeters

| Inches | Millimeters | Inches | Millimeters |
|--------|-------------|--------|-------------|
| 0.001 | 0.0254 | 0.2 | 5.080 |
| 0.002 | 0.0508 | 0.3 | 7.62 |
| 0.003 | 0.0762 | 0.4 | 10.16 |
| 0.004 | 0.1016 | 0.5 | 12.70 |
| 0.005 | 0.1270 | 0.6 | 15.24 |
| 0.006 | 0.1524 | 0.7 | 17.78 |
| 0.007 | 0.1778 | 0.8 | 20.32 |
| 0.008 | 0.2032 | 0.9 | 22.86 |
| 0.009 | 0.2286 | 1 | 25.40 |
| 0.01 | 0.2540 | 2 | 50.80 |
| 0.02 | 0.5080 | 3 | 76.20 |
| 0.03 | 0.7620 | 4 | 101.6 |
| 0.04 | 1.016 | 5 | 127.0 |
| 0.05 | 1.270 | 6 | 152.4 |
| 0.06 | 1.524 | 7 | 177.8 |
| 0.07 | 1.778 | 8 | 203.2 |
| 0.08 | 2.032 | 9 | 228.6 |
| 0.09 | 2.286 | 10 | 254.0 |
| 0.1 | 2.540 | | |

**Application Example:**

Using the table, convert 1.35 inches to millimeters.

Procedure: 1.35 inches can be writen as:

$1 + 0.3 + 0.05$ inches.

Converting each member of the addition using the table we find:

25.40 + 7.62 + 0.270 =33.29 mm

## TABLE 60
## Millimeters to Decimal Inches

| Millimeters | Decimal Inches | Millimeters | Decimal Inches |
|---|---|---|---|
| 0.01 | 0.000394 | 0.6 | 0.02358 |
| 0.02 | 0.000786 | 0.7 | 0.02751 |
| 0.03 | 0.001179 | 0.8 | 0.03144 |
| 0.04 | 0.001572 | 0.9 | 0.03537 |
| 0.05 | 0.001965 | 1 | 0.0389 |
| 0.06 | 0.002358 | 2 | 0.0786 |
| 0.07 | 0.002751 | 3 | 0.1179 |
| 0.08 | 0.003144 | 4 | 0.1572 |
| 0.09 | 0.003537 | 5 | 0.1965 |
| 0.1 | 0.003937 | 6 | 0.2358 |
| 0.2 | 0.00786 | 7 | 0.2751 |
| 0.3 | 0.01179 | 8 | 0.3144 |
| 0.4 | 0.01572 | 9 | 0.3537 |
| 0.5 | 0.01965 | 10 | 0.3890 |

**Application Example:**

Using the table, convert 2.47 mm to inches:

Procedure: 2.47 mm can be decomposed as follows:

$$I = 2 + 0.4 + 0.07$$

Converting each member by the table:

I = 0.0786 + 0.01572 + 0.002751

I = 0.097051 inches

# TEMPERATURE CONVERSIONS

## 192. DEGREES CELSIUS (C), DEGREES FAHRENHEIT (F), DEGREES KELVIN (K) AND DEGREES REAMUR (R)

The next formulas are used in temperature conversions. The scale Celsius is also called Centigrade and the scale Kelvin is also called Absolute. *Figure 172* shows the reference points in all scales (Water boiling point and water freezing point).

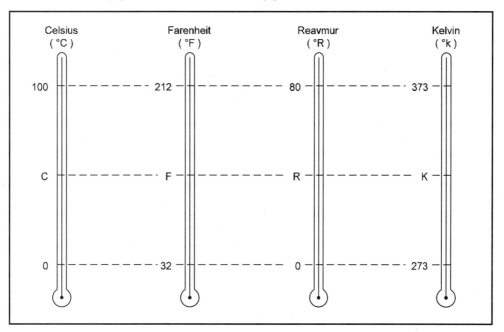

*Figure 172*

**Formula 192.1**

**Celsius to Fahrenheit**

$$F = \left(\frac{9}{5} xC\right) + 32$$

## Formula 192.2
## Fahrenheit to Celsius

$$C = \frac{5}{9}x(F - 32)$$

Where:

F is the temperature in degrees Fahrenheit (F)

C is the temperature in degrees Celsius (C)

**Application Example:**

Convert 75°F to Celsius.

Data:

F = 75 degrees

C = ?

Applying Formula 192.2:

$$C = \frac{5}{9}x(75 - 32)$$
$$C = \frac{5}{9}x43$$
$$C = 23.88$$

The correspondent temperature is 23.88° C.

## Formula 192.3
## Celsius to Kelvin

$$K = C + 273.16$$

## Formula 192.4
## Kelvin to Celsius

$$C = K - 273.16$$

Where:

> K is the temperature in degrees Kelvin (K)
>
> C is the temperature in degrees Celsius (C)

## Formula 192.5
## Fahrenheit to Kelvin

$$K = \left[\frac{5}{9}(F - 32)\right] + 273.16$$

**Application Example:**

Determine the temperature in Kelvin equivalent to 40 degrees Fahrenheit.

Data:

> F = 40°F
>
> C = ?

Using Formula 192.5:

$$K = \left[\frac{5}{9} x(40 - 32)\right] + 273.16$$

$$K = \left[\frac{5}{9} x8\right] + 273.16$$

$$K = 4.44 + 273.16 = 277.60$$

The correspondent temperature is 277.60° K.

## Formula 192.6
## Kelvin to Fahrenheit

$$F = \left[\frac{9}{5}(K - 273.16)\right] + 32$$

Where:

K is the temperature in degrees Kelvin (K)

F is the temperature in degrees Fahrenheit (F)

## Formula 192.7
## Celsius to Reamur

$$C = \text{Re} \, x \frac{5}{4}$$

## Formula 192.8
## Reamur to Celsius

$$\text{Re} = Cx \frac{4}{5}$$

Where:

Re is the temperature in degrees Reamur (Re)

C is the temperature in degrees Celsius (C)

## Formula 192.9
## Fahrenheit to Reamur

$$\text{Re} = \frac{4}{9}(F - 32)$$

## Formula 192.10
## Reamur to Fahrenheit

$$F = \left( \frac{9}{4} x \text{Re} \right) + 32$$

Where:

F is the temperature in degrees Fahrenheit (F)

Re is the temperature in degrees Reamur (Re)

## Formula 192.11
## Kelvin to Reamur

$$Re = \frac{4}{5}(K - 273.16)$$

## Formula 192.12
## Reamur to Kelvin

$$K = \left(\frac{5}{4} x Re\right) + 273.16$$

Where:

Re is the temperature in degrees Reamur (Re)

K is the temperature in degrees Kelvin (K)

## TABLE 61
## Degrees Celsius to Fahrenheit Conversion

| C | F | C | F | C | F |
|------|------|------|------|------|------|
| -50 | -58 | 20 | 68 | 90 | 194 |
| -45 | -49 | 25 | 77 | 95 | 203 |
| -40 | -40 | 30 | 86 | 100 | 212 |
| -35 | -31 | 35 | 93 | 105 | 221 |
| -30 | -22 | 40 | 104 | 110 | 230 |
| -25 | -13 | 45 | 113 | 115 | 239 |
| -20 | -4 | 50 | 122 | 120 | 248 |
| -15 | 5 | 55 | 131 | 125 | 257 |
| -10 | 14 | 60 | 140 | 130 | 266 |
| -5 | 23 | 65 | 149 | 135 | 275 |
| 0 | 32 | 70 | 158 | 140 | 284 |
| 5 | 41 | 75 | 167 | 145 | 293 |
| 10 | 50 | 80 | 176 | 150 | 302 |
| 15 | 59 | 85 | 185 | | |

## TABLE 62

## Farenheit to Celsius Conversion

| F | C | F | C | F | C |
|---|---|---|---|---|---|
| -50 | -46 | 50 | 10 | 125 | 52 |
| -40 | -40 | 55 | 13 | 130 | 54 |
| -30 | -34 | 60 | 16 | 135 | 57 |
| -20 | -29 | 65 | 18 | 140 | 60 |
| -10 | -23 | 70 | 21 | 145 | 63 |
| 0 | -18 | 75 | 24 | 150 | 66 |
| 5 | -15 | 80 | 27 | 155 | 68 |
| 10 | -12 | 85 | 29 | 160 | 71 |
| 15 | -9 | 90 | 32 | 165 | 74 |
| 20 | -7 | 95 | 35 | 170 | 77 |
| 25 | -4 | 100 | 38 | 175 | 79 |
| 30 | -1 | 105 | 41 | 180 | 82 |
| 35 | 1.7 | 110 | 43 | 185 | 85 |
| 40 | 4 | 115 | 46 | 190 | 88 |
| 45 | 7 | 120 | 49 | 195 | 91 |
|  |  |  |  | 200 | 93 |

# SOUND

# 193. SOUND WAVELENGTH

The velocity of sound waves advances is called *wave velocity*. The velocity depends on the properties of the medium and in some cases on the frequency. The dependence of the wave velocity on the frequency is called *dispersion* of the velocity.

The wavelength is defined as the distance travelled by the wave in one period. In consequence, the wavelength depends on the velocity of the wave in the considered medium.

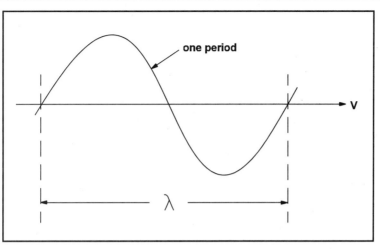

*Figure 173*

## Formula 193.1
## Sound Wavelength

$$\lambda = \frac{v}{f}$$

Where:

$\lambda$ is the wavelength in meters (m)

v is the propagation velocity in meters per second (m/s)

f is the frequency in hertz (Hz)

NOTE: If the velocity (v) is given in feet/second the wavelength is found in feet.

### Application Example:

Calculate the wavelength of a sound wave with 3000 Hz of frequency, given the velocity of sound 340 m/s.

Data:

f = 3 000 Hz

v = 340 m/s

$\lambda$ = ?

Using Formula 193.1:

$$\lambda = \frac{340}{3000} = 0.113 \text{ m or } 11.3 \text{ cm}$$

## Derivated Formula:
## Wavelength x Period

$$\lambda = vxT$$

Where:

$\lambda$ is the wavelength in meters (m)

v is the velocity of propagation in meters per second (m/s)

T is the period in seconds (s)

(See Table 17 for more information about sound velocities in solid materials)

# 194. VELOCITY OF LONGITUDINAL SOUND WAVES

The velocity of a longitudinal acoustic wave in a solid material depends on the nature of the material and can be calculated by the next formula:

## Formula 194.1
## Mechanical Longitudinal Waves in a Solid

$$vL = \sqrt{\frac{E}{\rho}}$$

Where:

VL is the velocitiy of the longitudinal waves in meters per second (m/s)

E is the Young's Modulus (*)

$\rho$ is the density

(*) Young's Modulus is numerically equal to the stress required to double the length of a body. Actually, in most bodies, rupture occurs at considerable smaller stresses.

# TABLE 63
## Young's Modulus for some Solid Materials

| Material | Young's Modulus (kg/mm²) |
|---|---|
| Aluminum, rolled | 6 900 |
| Aluminum wire | 7 000 |
| Bakelite | 200-300 |
| Brass, rolled | 10 000 |
| Cast iron | 11 000 - 16 000 |
| Celluloid | 1.7 to 2.0 |
| Constatan | 10 000 |
| Copper casting | 8 400 |
| Copper, rolled | 11 000 |
| Duralumin, rolled | 7 100 |
| Glass | 5 000 - 8 000 |
| Granite | 5 000 |
| Ice | 1 000 |
| Invar | 14 000 |
| Lead | 1 700 |
| Manganin | 12 600 |
| Marble | 4 000 - 6 000 |
| Phospher bronze, rolled | 11 500 |
| Plexiglass | 530 |
| Rubber | 0.8 |
| Steel alloy | 21 000 |
| Steel carbon | 19 000 - 21 000 |
| Vinyl plastic | 300 |
| Wood | 300 - 1 600 |
| Zinc, rolled | 8 400 |

# 195. SOUND PRESSURE

The maximum increase in pressure in the medium due to the presence of a sound wave is called sound pressure. Pressure is the amount of force acting over a 1 cm unit area of surface. Sound pressure is normally expressed in dynes/cm² or microbar ( $\mu bar$ ). The next formulas describe relationships between sound presure and the maximum velocity of vibration of particles.

## Formula 195.1
## Sound Pressure x Vibration of Particles

$$p = vxz$$

Where:

P is the sound pressure in $\mu$ bar (*)

v is the velocity of vibration in centimeters per second (cm/s)

z is the specific acoustic resistance in g/cm²s

(*) 1 $\mu$ bar = 0.1 Pa = 1 g/cm².s or 1 bar = $10^6$ dynes/cm²

## Other Formulas:

## Formula 195.2
## Specific Acoustic Resistance

$$z = vx\rho x \cos\varphi$$

Where:

z is the specific acoustic resistance in g/cm²

v is the velocity of sound in cm/s

$\rho$ is the medium density in g/cm³

$\varphi$ is the phase angle between p and v

p is the acoustic pressure in $\mu$ bar

**Formula 195.3**

**Tangent of** $\varphi$

$$tg\varphi = \frac{\lambda}{2x\pi xr}$$

Where:

$\varphi$ is the phase angle between p and v

$\lambda$ is the wavelength in meters (m)

$\pi$ is 3.1416

r is the distance from the sound souce in meters (m)

# 196. SOUND PRESSURE LEVEL

A change in pressure (pressure level) can be expressed in decibels. Any sound pressure compared with the 0.0002 dyne/cm² reference is expressed in sound pressure level (SPL) and calculated by the next formula.

**Formula 196.1**

**Sound Pressure Level**

$$SPL = 20x\lg\frac{P}{Po}$$

*or*

$$SPL = 20x\lg(5000xP)$$

Where:

PL is the sound pressure level in dB

P is the sound pressure in $\mu$ bar

Po is the standard reference ($2x10^{-4}$ $\mu$ bar or 0.0002 dynes/cm² $= \dfrac{1}{5000}\mu bar$ )

# 197. SOUND INTENSITY

Sound intensities are measured in watts per square centimeter (w/cm²). The intensity is the amount of power transmitted along a wave through an area of 1 square centimeter at right angles to the direction of the propagation of the wave, as shown by *Figure 174*. The ratio between two sound intensities can be expressed as a decibel difference or an absolute value as compared to a reference, as shown by the next formulas:

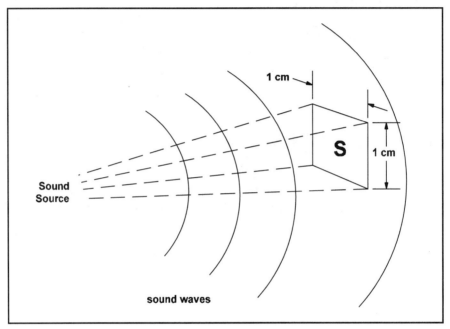

*Figure 174*

## Formula 197.1
## Sound Intensity (compared)

$$L = 10 \lg\left(\frac{I}{Io}\right)$$

Where:

L is the sound intensity ratio in dB

I, Io are the compared sound intensities in w/cm²

## Formula 197.2

## Sound Intensity Compared to a Reference (intensity level)

$$IL = 10\lg\left(\frac{I}{10^{-12}}\right)$$

Where:

IL is the intensity level in w/cm²

I is the sound intensity in w/cm²

$10^{-12}$ w/cm² is the reference sound level corresponding to the audible sounds

## TABLE 64

## Sound Intensity Levels

| Intensity (W/cm²) | Effect | Common Sources |
|---|---|---|
| $10^{-16}$ | Threshold of Hearing | Reference Level |
| $10^{-15}$ | Very Faint | Anechoic Room, Breathing at 1 ft |
| $10^{-14}$ | Very Faint | Very Quiet Residence |
| $10^{-13}$ | Faint | Average Whisper, Quiet Suburban Garden |
| $10^{-12}$ | Faint | Average Residence |
| $10^{-11}$ | Moderate | Church, Soft Violin Solo, Quiet Car |
| $10^{-10}$ | Moderate | Quiet Residential Street, Average Office |
| $10^{-9}$ | Loud | Busy Street, Noisy Saloon |
| $10^{-8}$ | Loud | Factory, Very Loud Radio, Loud Speech |
| $10^{-7}$ | Very Loud | Noisy Factory, Loudest Orchestral Music Heavy Street Traffic |
| $10^{-6}$ | Soft Discomfort | Loud Musical Peaks, Subway, Thunder |
| $10^{-5}$ | Deafening | Loud Bus Horn, Boiler Factory, Bass Drum |
| $10^{-4}$ | Heavy Discomfort | Pneumatic Hammer, Car Horn |
| $10^{-3}$ | Threshold of Pain | |
| $10^{-2}$ | Pain | Large Ship Siren |
| $10^{-1}$ | Impairs Hearing | Jet Engine |

# LOUDSPEAKER CROSSOVER NETWORKS

# 198.SIMPLE FOR TWEETER (6 db/octave)

The configuration shown in *Figure 175* is the simplest crossover network used to derive high-frequency audio signals to tweeters. The capacitors must be non-polarized types.

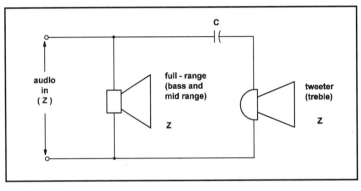

*Figure 175*

## Formula 198.1
## 6 dB/Octave — 2 Channels

$$C = \frac{2}{\pi x f x Z}$$

Where:

C is the capacitor in farads (F)

f is the crossover frequency in hertz (Hz)

Z is the loudspeaker impedance em ohms ( $\Omega$ )

$\pi$ is 3.1416

**Application Fxample:**

Determine the capacitance to be used in the crossover circuit shown by *Figure 175* when the loudspeakers are 8-ohm impedance types and the desired crossover frequency is 2 kHz.

Data: Z = 8 ohms

> f = 2 000 Hz
>
> C = ?

Using Formula 198.1:

$$C = \frac{2}{3.14 x 2000 x 8}$$

$$C = \frac{2}{50240}$$

$$C = 0.0000398 F$$

$$C = 39.8 \mu F$$

A 40 $\mu F$ non-polarized electrolytic capacitor can be used in this application.

## Derivated Formulas:

## Formula 198.2
## Capacitance in Microfarads

$$C = \frac{2 x 10^6}{\pi x f x Z}$$

Where:

> C is the capacitance in microfarads ( $\mu F$ )
>
> f is the crossover frequency in hertz (Hz)
>
> Z is the impedance in ohms ( $\Omega$ )
>
> $\pi$ is 3.1416

# 199. 6 dB/Octave CROSSOVER NETWORK-II

The next formulas are used to calculate the elements of the loudspeaker crossover network shown in *Figure 176*.

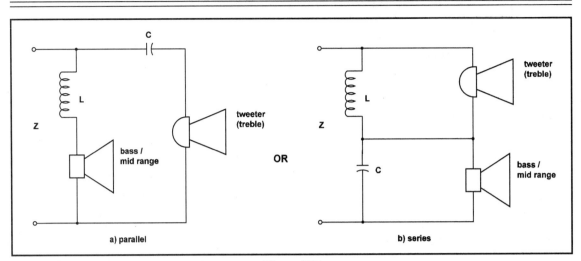

*Figure 176*

## Formula 199.1
## Capacitor

$$C = \frac{10^6}{2x\pi xfxZ}$$

Where:

    C is the capacitor in microfarads ($\mu$)

    f is the crossover frequency in hertz (Hz)

    Z is the impedance of the system in ohms ($\Omega$)

    $\pi$ is 3.1416

## Formula 199.2
## Inductance

$$L = \frac{Zx10^3}{2x\pi xf}$$

Where:

> L is the inductance in millihenrys (mH)
>
> Z is the impedance in ohms ($\Omega$)
>
> f is the crossover frequency in hertz (Hz)
>
> $\pi$ is 3.1416

# 200. TWO-CHANNEL 12 dB LOUDSPEAKER CROSSOVER

The next formulas are used to determine the elements of the circuit shown in *Figure 177*. The crossover frequency is determined according to the tweeter (high-frequency loudspeaker) characteristics given by the manufacturer.

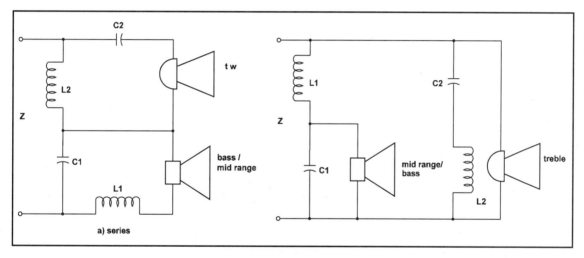

*Figure 177*

## Formula 200.1
## Capacitance C1

$$C1 = \frac{1.6x10^6}{2x\pi xfxZ}$$

**Formula 200.2**
**Capacitance C2**

$$C2 = \frac{10^6}{2x\pi xfxZ}$$

**Formula 200.3**
**Inductance L1**

$$L1 = \frac{Zx10^3}{2x\pi xf}$$

**Formula 200.4**
**Inductance L2**

$$L2 = \frac{Zx10^3}{3.2x\pi xf}$$

Where:

C1 and C2 are the capacitances in microfarads ($\mu F$)

L1 and L2 are the inductances in millihenrys (mH)

Z is the impedance of the system in ohms ($\Omega$)

f is the crossover frequency in hertz (Hz)

$\pi$ is 3.1416

# 201. TWO-CHANNEL 18 dB/octave PI-CROSSOVER NETWORK

The following formulas are valid when designing a two-loudspeaker system, such as the one in *Figure 178*.

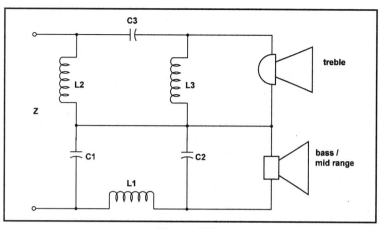

*Figure 178*

## Formula 201.1
## Capacitance C1

$$C1 = \frac{1.6x10^6}{2x\pi xfxZ}$$

## Formula 201.2
## Capacitance C2

$$C2 = \frac{10^6}{2x\pi xfxZ}$$

## Formula 201.3
## Capacitance C3:

$$C3 = \frac{10^6}{4x\pi xfxZ}$$

Where:

C1, C2 and C3 are the capacitances in microfarads ( $\mu F$ )

f is the crossover frequency in hertz (Hz)

Z is the impedance of the system in ohms ( $\Omega$ )

$\pi$ is 3.1416

**Formula 201.4**

**Inductance L1**

$$L1 = \frac{Zx10^3}{\pi x f}$$

**Formula 201.5**

**Inductance L2**

$$L2 = \frac{Zx10^3}{3.2 x \pi x f}$$

**Formula 201.6**

**Inductance L3**

$$L3 = \frac{Zx10^3}{2 x \pi x f}$$

Where:

  L1, L2 and L3 are the inductances in millihenrys (mH)

  Z is the impedance of the system in ohms ($\Omega$)

  f is the frequency in hertz (Hz)

  $\pi$ is 3.1416

# 202. TWO-CHANNEL 18 dB/Octave T-CROSSOVER NETWORK

The next formulas are valid when calculating elements for the circuit shown in *Figure 179*.

**Formula 202.1**

**Capacitance C1**

$$C1 = \frac{10^6}{\pi x f x Z}$$

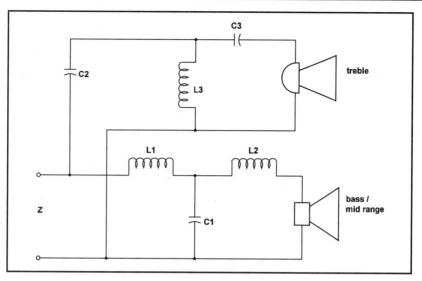

*Figure 179*

## Formula 202.2
## Capacitance C2

$$C2 = \frac{10^6}{3.2 \, x\pi x f x Z}$$

## Formula 202.3
## Capacitance C3

$$C3 = \frac{10^6}{2 \, x\pi x f x Z}$$

Where:

C1, C2 and C3 are the capacitances in microfarads ( $\mu F$ )

f is the frequency in hertz (Hz)

Z is the impedance of the system in ohms ( $\Omega$ )

$\pi$ is 3.1416

## Formula 202.4
## Inductance L1

$$L1 = \frac{Zx10^3}{\pi xf}$$

## Formula 202.5
## Inductance L2

$$L2 = \frac{Zx10^3}{3.2x\pi xf}$$

## Formula 202.6
## Inductance L3

$$L3 = \frac{Zx10^3}{4x\pi xf}$$

Where:

L1, L2 and L3 are the inductances in millihenrys (mH)

Z is the impedance of the system in ohms ($\Omega$ )

f is the frequency in hertz (Hz)

$\pi$ is 3,1416

# 203. CROSSOVER AND DESIGN FREQUENCIES
# FOR 3-WAY NETWORKS

As shown in *Figure 180*, when designing a three-channel crossover network, you must use four frequencies in the calculations. These frequencies are used in the practical projects.

## Formula 203.1
## Frequency Band Ratio and Design Ratio

$$\frac{f_4}{f_3} = \frac{f_2}{f_1} - 1$$

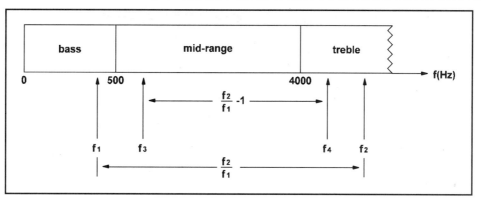

*Figure 180*

## Formula 203.2
## Calculating f3

$$f_3 = \sqrt{\dfrac{f_1 x f_2}{\dfrac{f_2}{f_1} - 1}}$$

## Formula 203.4
## Calculating f4

$$f_4 = f_3 x \left( \dfrac{f_2}{f_1} - 1 \right)$$

Where:

f1 is the lower crossover frequency in hertz (Hz)

f2 is the upper crossover frequency in hertz (Hz)

f3 is the lower design frequency in hertz (Hz)

f4 is the upper design frequency in hertz (Hz)

f2/f1 is the frequency band ratio

f4/f3 is the design ratio

In typical calculations f1 is about 500 Hz and f2 4000 Hz. These correspond to the transition points from bass to midrange and from midrange to treble.

# 204. THREE-CHANNEL 6dB/Octave CROSSOVER NETWORK (Series)

The following formulas are used to calculate inductances and capacitances in the series circuit shown in *Figure 181*.

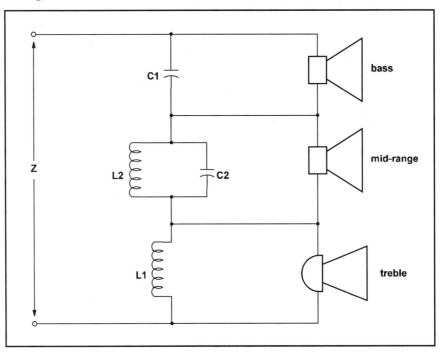

*Figure 181*

## Formula 204.1
### Capacitance C1

$$C1 = \frac{1}{2x\pi xf_2 xZ}$$

## Formula 204.2
### Capacitance C2

$$C2 = \frac{1}{2x\pi xf_4 xZ}$$

## Formula 204.3
## Inductance L1

$$L1 = \frac{Z}{2x\pi x f_1}$$

## Formula 204.4
## Inductance L2

$$L2 = \frac{Z}{2x\pi x f_3}$$

Where:

C1, C2 are the capacitances in microfarads ($\mu F$)

L1 and L2 are the inductances in millihenrys (mH)

f1, f2, f3 and f4 are the frequencies calculated by Formula 203 in hertz (Hz)

$\pi$ is 3.1416

Z is the impedance of the system in ohms ($\Omega$)

# 205. THREE-CHANNEL 6 dB/Octave CROSSOVER (Parallel)

The next formulas are used with the formulas of 203 to calculate the elements of the parallel crossover network shown in *Figure 182*.

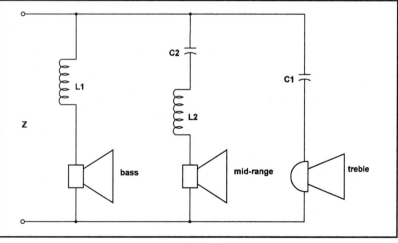

*Figure 182*

## Formula 205.1
## Capacitance C1

$$C1 = \frac{1}{2 x \pi x f_2 x Z}$$

## Formula 205.2
## Capacitance C2

$$C2 = \frac{1}{2 x \pi x f_3 x Z}$$

## Figure 205.3
## Inductance L1

$$L1 = \frac{Z}{2 x \pi x f_1}$$

## Formula 205.4
## Inductance L2

$$L2 = \frac{Z}{2 x \pi x f_4}$$

Where:

C1 and C2 are the capacitances in microfarads ($\mu F$)

L1 and L2 are the inductances in millihenrys (mH)

f1, f2, f3 and f4 are the frequencies determined by the #204 formulas, in hertz (Hz)

Z is the impedance of the system in ohms ($\Omega$)

$\pi$ is 3.1416

# 206. THREE-CHANNEL 12 dB/octave CROSSOVER (Series)

The formulas below are valid when designing a crossover network like the one shown in *Figure 183*. The frequencies f1 to f4 are the ones determined by Formulas 204x.

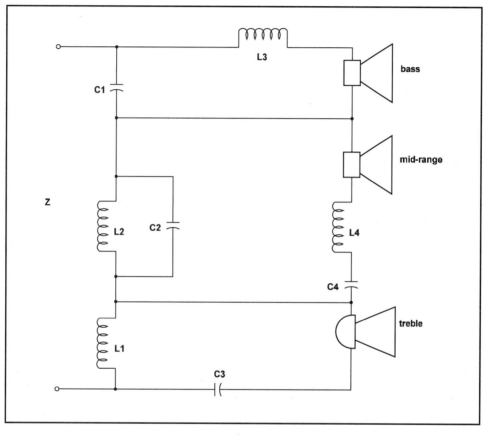

*Figure 183*

## Formula 206.1
## Capacitance C1

$$C1 = \frac{\sqrt{2}}{2 x \pi x f_1 x Z}$$

**Formula 206.2**
**Capacitance C2**

$$C2 = \frac{\sqrt{2}}{2x\pi x f_4 x Z}$$

**Formula 206.3**
**Capacitance C3**

$$C3 = \frac{\sqrt{2}}{2x\pi x f_3 x Z}$$

**Formula 206.4**
**Capacitance C4**

$$C4 = \frac{\sqrt{2}}{2x\pi x f_3 x Z}$$

**Formula 206.5**
**Inductance L1**

$$L1 = \frac{Z}{2x\sqrt{2}x\pi x f_1}$$

**Formula 206.6**
**Inductance L2**

$$L2 = \frac{Z}{2x\sqrt{2}x\pi x f_3}$$

**Formula 206.7**
**Inductance L3**

$$L3 = \frac{Z}{2x\sqrt{2}x\pi x f_1}$$

## Formula 206.8
## Inductance L4

$$L4 = \frac{Z}{2x\sqrt{2}x\pi x f_4}$$

Where:

Z is the impedance of the system in ohms ( $\Omega$ )

C1, C2, C3 and C4 are the capacitances in microfarads ( $\mu F$ )

L1, L2, L3 and L4 are the inductances in millihenrys (mH)

f1 to f4 are the frequencies calculated according Formulas #203, in hertz (Hz)

$\pi$ is 3.1416

# 207. THREE-CHANNEL 12 dB/Octave CROSSOVER (Parallel)

The formulas below are valid when calculating the elements for the circuit of *Figure 184*.

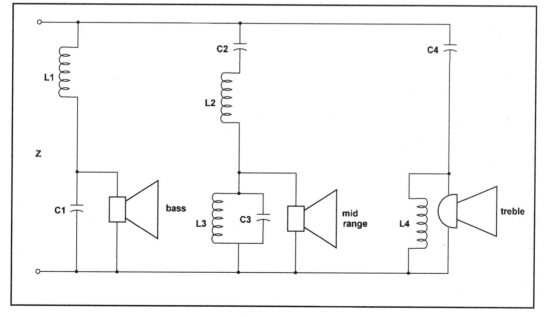

*Figure 184*

**Formula 207.1**
**Capacitance C1**

$$C1 = \frac{1}{2x\sqrt{2}x\pi x f_1 xZ}$$

**Formula 207.2**
**Capacitance C2**

$$C2 = \frac{1}{2x\sqrt{2}x\pi x f_3 xZ}$$

**Formula 207.3**
**Capacitance C3**

$$C3 = \frac{1}{2x\sqrt{2}x\pi x f_4 xZ}$$

**Formula 207.4**
**Capacitance C4**

$$C4 = \frac{1}{2x\sqrt{2}x\pi x f_2 xZ}$$

**Formula 207.5**
**Inductance L1**

$$L1 = \frac{Zx\sqrt{2}}{2x\pi x f_1}$$

**Formula 207.6**
**Inductance L2**

$$L2 = \frac{Zx\sqrt{2}}{2x\pi x f_4}$$

**Formula 207.7**
**Inductance L3**

$$L3 = \frac{Zx\sqrt{2}}{2x\pi x f_3}$$

**Formula 207.8**
**Inductance L4**

$$L4 = \frac{Zx\sqrt{2}}{2x\pi x f_2}$$

Where:

Z is the impedance of the system in ohms ($\Omega$)

C1, C2, C3 and C4 are the capacitances in microfarads ($\mu F$)

L1, L2, L3 and L4 are the inductances in millihenrys (mH)

f1 to f4 are the frequencies calculated by Formulas #203, in hertz (Hz)

$\pi$ is 3.1416

# 208. WEELER'S FORMULA

When calculating inductances to be used in crossover networks in the range of millihenrys (mH), a useful formula is the one created by Weeler. But this formula only gives good results if the dimensions **a**, **b** and **c** have the same order of magnitude, as shown in *Figure 185*.

**Formula 208.1**
**Weeler's Formula**

$$L = \frac{0.315xa^2xn^2}{6a + 9b + 10c}$$

Where:

L is the inductance in microhenrys ($\mu H$)

a,b and c are the dimensions of the coil in centimeters (cm)

n is the number of turns

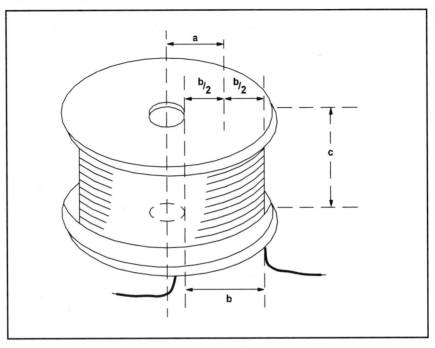

*Figure 185*

## Derivated Formula:

## Formula 208.2
## Number of Turns

$$n = \sqrt{\frac{Lx(6a + 9b + 10c)}{0.315xa^2}}$$

Where:

    n is the number of turns

    L is the inductance in microhenrys ( $\mu H$ )

    a,b and c are the dimensions of the coil in centimeters (cm)

### Application Example:

How many turns of wire we must wind to have a 5 mH coil when the dimensions a, b and c are respectively: 3, 2 and 2 cm?

Data:

$L = 5$ mH $= 5\,000$ $\mu H$

$a = 3$ cm

$b = 2$ cm

$c = 2$ cm

$n = ?$

Using Formula 208.2:

$$n = \sqrt{\frac{5000x(6x3 + 9x2 + 10)}{0.315x3^2}}$$

$$n = \sqrt{\frac{5000x(18 + 18 + 20)}{0.315x9}}$$

$$n = \sqrt{\frac{5000x56}{2.835}}$$

$$n = \sqrt{\frac{280000}{2.835}}$$

$$n = \sqrt{98765}$$

$$n = 314$$

The number of turns is 314.

The gauge of the used wire must now be determined according to the space to be filled in the form and also the power to be managed by the coil.

# Part 6

# Optoelectronics

# OPTOELECTRONICS

The following sequence of formulas, tables and calculations deal with devices operating with light. TV and sensors will also be approached.

## TABLE 65
### Optical Radiation and Light

UV = Ultraviolet, IR = Infra-red

| Wavelength (nm) | Designation |
|---|---|
| 100 - 280 | UV - C |
| 280 - 315 | UV - B |
| 315 - 380 | UV - A |
| 380 - 440 | Visible - violet |
| 440 - 495 | Visible - blue |
| 495 - 580 | Visible - green |
| 580 - 640 | Visible - yellow |
| 640 - 750 | Visible - red |
| 750 - 1 400 | IR - A |
| 1400 - 3000 | IR - B |
| 3000 - 1,000,000 | IR - C |

*Figure 186*

# 209. WAVELENGTH AND FREQUENCY FOR LIGHT RADIATION

Although the general formula for electromagnetic waves was given at the beginning of this book, we are repeating it now in a specific mode (with calculations) for light radiation (including IR and UV).

# Formula 209.1
## Wavelength vs. Frequency

$$\lambda = \frac{c}{f}$$

Where:

$\lambda$ is the wavelength in meters (m)

c is the speed of light in vacuum = 300,000,000 m/s (*)

f is the frequency in hertz (Hz)

(*) The exact value is 2.99792 x $10^8$ m/s but in general calculations we use 3 x $10^8$ m/s.

## Alternative Formula:

# Formula 209.2
## Wavelength vs. Frequency

$$\lambda = \frac{300000}{f}$$

Where:

$\lambda$ is found in nanometers (nm)

f is the frequency in terahertz (THz)

## Derivated Formulas:

# Formula 209.3
## Frequency x Wavelength

$$f = \frac{c}{\lambda}$$

Where:

f is the frequency in hertz (Hz)

c is the speed of light in vaccum (300,000,000 m/s)

$\lambda$ is the wavelength in nanometers (nm)

NOTE: In optics we also use as a unit of wavelength the Angstrom (Å).

$$1 \text{ Angstrom} = 10^{-1} \text{ nm}$$

**Application Example:**

What is the frequency of a 600 nm radiation?

Data:

$c = 300\ 000$ km/s

$\lambda = 600$ nm

$f = ?$

Using Formula 209.3:

$$f = \frac{300000}{600}$$
$$f = 500 THz$$

## TABLE 66
## Photometric Conversion Table

| Unit | Equivalence |
|------|-------------|
| 1 candela/square inch | 452 footlamberts, 0.487 lamberts |
| 1 lambert | 929 footlamberts, 2.054 candela/sq in., 0.318 candela/sq in. |
| 1 millilambert | 0.01 lambert, 0.929 footlambert, 0.002 candela/sq. in. |
| 1 lumen | $1.496 \times 10^{-3}$ watts, 0.07958 spherical candel/power |
| lumen/square feet | 1 footcandle, 10.76 lumens/square meter |
| 1 lux | 0.0929 footcandles |
| 1 angstron | $10^{-10}$ meter, $10^{-4}$ microns |

# 210. LUMINOUS INTENSITY AND FLUX

The energy radiated by a body per second is the intensity of radiation, or luminous intensity, and the energy transmitted by a light wave per second to a surface is called the flux of radiation, or luminous flux through the surface ( $\Phi$ ).

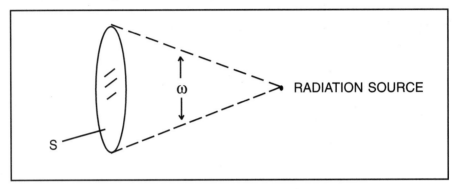

*Figure  187*

## Formula 210.1
## Luminous flux & Luminous intensity

$$\Phi = Ix\omega$$

Where:

$\Phi$ is the luminous flux in lumens (lm)

I is the luminous intensity in candelas (cd)

$\omega$ is the solid angle element covered by the radiation in steraradians (sr)

## Derivated formulas:

## Formula 210.2
## Luminous Flux

$$\Phi = Ix4x\pi xsin^2\frac{\alpha}{4}$$

Where:

    $\Phi$ is the luminous flux in lumens (lm)

    I is the luminous intensity in candelas (cd)

    $\alpha$ is the angular overture in sterarians (sr)

    $\pi$ is 3.1416

## Formula 210.3
## Luminous Intensity

$$I = \frac{\Phi}{\omega}$$

Where:

    I is the luminous intensity in candelas (cd)

    $\Phi$ is the luminous flux in lumens (lm)

    $\omega$ is the solid angle element covered by the radiation in steradians (sr)

# 211. LUMINOUS DENSITY

Is defined as the amount of luminous intensity by unit of area. Two formulas are used to calculate this quantity, depending on the incidence angle of radiation as shown in *Figure 188*.

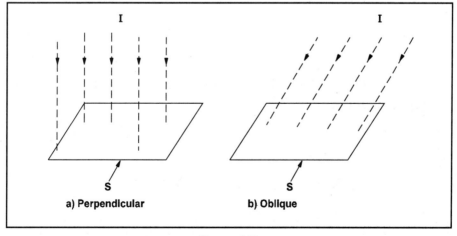

*Figure 188*

## Formula 211.1
## Perpendicular

$$L = \frac{I}{S}$$

*or*

$$L = \frac{\Phi}{\omega x S}$$

Where:

L is the luminous density of a reflective surface or luminiscent in candelas per square centimeters (cd/cm$^2$)

I is the luminous intensity in candelas (cd)

$\Phi$ is the luminous flux in lumens (lm)

S is the surface area in square centimeters (cm$^2$)

$\omega$ is the solid angle in stereradians (sr)

## Formula 211.2
## Obliquous

$$L = \frac{I}{S x \cos\varepsilon}$$

*or*

$$L = \frac{\Phi}{\omega x S x \cos\varepsilon}$$

Where:

L is the luminous density of a luminescent surface or reflective surface in candelas per square centimeter (cd/cm$^2$)

I is the luminous intensity in candelas (cd)

S is the surface area in square centimeters (cm$^2$)

$\Phi$ is the luminous flux in lumens (lm)

$\omega$ is the solid angle in stereradians (sr)

$\varepsilon$ is the inclination angle referred to the perpendicular line in degrees

## TABLE 67

## Radiometric and Photometric Parameters

| | Parameter | Symbol | Dimension | Unit |
|---|---|---|---|---|
| radiometric | radiant intensity | Ie | $\dfrac{power}{solid.angle}$ | $Wsr^{-1}$ |
| photometric | luminous intensity | Iv | $\dfrac{power}{solid.angle}$ | $Imsr^{-1}$, cd |
| radiometric | radiance | Le | $\dfrac{power}{area.of.active.region x solid.angle}$ | $\dfrac{W}{m^2 xsr}$ |
| photometric | luminance | Lv | $\dfrac{power}{area.of.active.reginx olid.angle}$ | $\dfrac{Candela}{m^2}$ |
| radiometric | radiant flux radiant power | Φ (Po) | power | W |
| photometric | luminous power Luminous flux | Φ (v) | Power | lm |
| radiometric | radiant energy | We | power x time | Ws |
| photometric | quantity of light (Luminous energy) | Wv | power x time | lms |
| radiometric | radiant exitance (radiant emittance) | Me | $\dfrac{power}{area.of.active.regim}$ | $W.m^{-2}$ |
| photometric | luminous exitance (luminous emittance) | Mv | $\dfrac{power}{area.of.active.region}$ | $lm.m^{-2}$ |
| radiometric | irradiance | Ee | $\dfrac{power}{area}$ | $W.m^{-2}$ |
| photmetric | illuminance | Ev | $\dfrac{power}{area}$ | $\dfrac{foot.candela.lm}{mm^2}$ |
| radiometric | radiant exposure (irradiation) | He | $\dfrac{energy}{area}$ | $Wsm^{-2}$ |
| photometric | light exposure | Hv | $\dfrac{luminous.enery}{area}$ | $\dfrac{lm.s}{m^2}; lx.s$ |

# TABLE 68
## Spectral Units

| Parameter | Formula | Unit |
|-----------|---------|------|
| Spectral irradiation | $He, \lambda = \dfrac{dHe}{d\lambda}$ | Wsm$^{-2}$ (nm$^{-1}$) |
| Spectral radiance | $Le, \lambda = \dfrac{dLe}{d\lambda}$ | Wm$^{-2}$ sr$^{-1}$ (nm$^{-1}$) |
| Spectral radiant power | $\Phi e, \lambda = \dfrac{d\Phi e}{d\lambda}$ | W (nm$^{-1}$) |
| Spectral radiant power | $\Phi e, v = \dfrac{d\Phi e}{dv}$ | W(Hz$^{-1}$) |
| Spectral radiant intensity | $Ie, \lambda = \dfrac{dIe}{d\lambda}$ | Wsr$^{-1}$(nm$^{-1}$) |
| Spectral radiant emittance | $Me.\lambda = \dfrac{dMe}{d\lambda}$ | Wm$^{-2}$(nm$^{-1}$) |
| Spectral irradiance | $Ee, \lambda = \dfrac{dEe}{d\lambda}$ | Wm$^{-2}$ (nm$^{-1}$) |

# TABLE 69
## Photopic Sensitivity of the Eye

This table shows the photopic sensitivity of the human eye $V(\lambda)$ and the photometric radiation equivalent $K(\lambda)$ as function of the wavelength $(\lambda)$ -

| Wavelength(nm) | Spectral Brightness Sensitivity of the Eye(Lv=10 cd/m²) | Photometric Radiation Equivalent (lm/W) |
|---|---|---|
| 380 | 0.00004 | 0.0272 |
| 400 | 0.0004 | 0.272 |
| 410 | 0.0012 | 0.816 |
| 420 | 0.0040 | 2.72 |
| | (continued next page) | |

| Wavelength(nm) | Spectral Brightness Sensitivity of the Eye(Lv=10 cd/m²) | Photometric Radiation Equivalent (lm/W) |
|---|---|---|
| 430 | 0.0116 | 7.89 |
| 440 | 0.023 | 15.64 |
| 450 | 0.038 | 25.84 |
| 460 | 0.060 | 40.8 |
| 470 | 0.091 | 61.88 |
| 480 | 0.139 | 94.52 |
| 490 | 0.208 | 141.44 |
| 500 | 0.323 | 219.64 |
| 510 | 0.503 | 342.04 |
| 520 | 0.710 | 482.8 |
| 530 | 0.862 | 586.16 |
| 540 | 0.954 | 648.72 |
| 550 | 1 | 680 |
| 560 | 0.995 | 676.6 |
| 570 | 0.952 | 647.36 |
| 580 | 0.870 | 591.6 |
| 590 | 0.757 | 514.76 |
| 600 | 0.631 | 429.08 |
| 610 | 0.503 | 342.04 |
| 620 | 0.381 | 259.08 |
| 630 | 0.265 | 180.2 |
| 640 | 0.175 | 119 |
| 650 | 0.107 | 72.76 |
| 660 | 0.061 | 41.48 |
| 670 | 0.032 | 21.76 |
| 680 | 0.017 | 11.56 |
| 690 | 0.0082 | 5.576 |
| 700 | 0.0041 | 2.788 |
| 710 | 0.0021 | 1.428 |
| 720 | 0.00105 | 0.714 |
| 730 | 0.00052 | 0.3536 |
| 740 | 0.00025 | 0.0170 |
| 750 | 0.00012 | 0.0816 |
| 760 | 0.00006 | 0.0408 |

## TABLE 70

## Characteristic of Some Common Light Sources

| Source | Characteristic | Value |
|---|---|---|
| Sun (noon) | Luminance | $4.7 \times 10^8$ foot lamberts |
| | Illuminance | $10^5$ lumensm/m$^2$ |
| Lightning Flash | Luminance | $2 \times 10^{10}$ foot lamberts |
| 100 W incandescent lamp | Total emission | 1630 photometric lumens |
| | | 30 radiometric watts |
| | Luminance | $2.6 \times 10^6$ foot lamberts |
| 40 W fluorescent lamp | Total emission | 2560 photometric lumens |
| | | 16 radiometric watts |
| | Luminance | 2000 foot lamberts |
| Moon | Luminance | 730 foot lamberts |
| | Illuminance | 0.27 photmetric lumens/m$^2$ |
| Starlight - magnitude zero | Illuminance | $2.6 \times 10^{-6}$ photometric lumens/m$^2$ |
| Starlight - 6$^{th}$ magnitude | Illuminance | $10^{-8}$ photometric lumens/m$^2$ |

# 212. LEDs

The current-limiting resistor of an LED DC circuit as shown in *Figure 189* is calculated by:

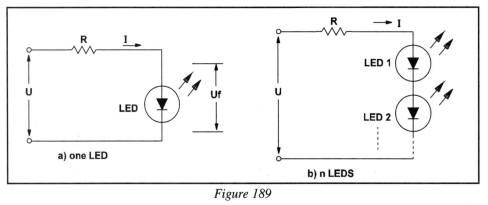

*Figure 189*

# Formula 212.1
## One LED

$$R = \frac{U - Uf}{I}$$

Where:

R is the resistance in ohms ($\Omega$)

U is the power supply voltage in volts (V)

Uf is the forward voltage fall in the LED (depends the type, see the table) in volts (V)

I is the current in amperes (A)

# Formula 212.2
## Two or more LEDs:

$$R = \frac{U - nxUf}{I}$$

Where:

R is the resistance in ohms ($\Omega$)

n is the number of LEDs

Other parameters as in previous formula

**Important:  nxUf<U**

# TABLE 71
## Forward Voltage Fall (Uf) in LEDs

| LED | Uf (V) |
|---|---|
| Red | 1.6 |
| Yellow/Orange | 1.8 |
| Green | 2.1 |
| Blue | 2.4 |

**Application Example:**

Determine value of R to be wired in series with a red LED when powering the circuit from a 12V power supply and draining a 20 mA current.

Data:

U = 12V

I = 20 mA = 0.02A

R = ?

Uf = 1.6V  (red)

Using Formula  212.1:

$$R = \frac{12 - 1.6}{0.02}$$

$$R = \frac{11.4}{0.02} = 570$$

R = 570 ohms

# 213. PHOTOCELL

The next formulas describe the principal properties of photo elements—solar cells, or lightcells as they are also called. The equivalent circuit with the parameters used by the formulas are shown in the figure below.

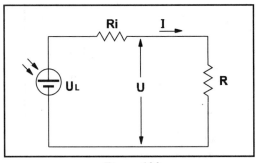

*Figure 190*

# Formula 213.1
# Photoelectric Voltage

$$U = \frac{U_L x R}{Ri + R}$$

Where:

U is the photoelectric voltage under illumination in volts (V)

$U_L$ is the photoelectric voltage in open circuit in volts (V)

R is the load resistance in ohms ($\Omega$ )

Ri is internal resistance of the photoelement in ohms ($\Omega$)

# Formula 213.2
# Sensitivity to Luminous Flux

$$s = \frac{If}{\Phi}$$

*or*

$$s = \frac{e}{S}$$

Where:

s is the sensitivity to the luminous flux in milliamperes per lumen (mA/lm)

If is photoelectric current in milliamperes (mA)

$\Phi$ is the luminous flux in lumens (lm)

e is luminous sensitivity in milliamperes per lux (ma/lx)

S is the area of the light-sensitive surface in square centimeters (cm$^2$ )

# Formula 213.3
# Ilumination Sensitivity

$$e = \frac{If}{E}$$

*or*

$$e = sxS$$

Where:

> e is the ilumination sensitivity in milliamperes per lux (mA/lx)
>
> If is the photoelectric current in milliamperes (mA)
>
> E is ilumination intensity in lux (lx)
>
> s is the sensitivity to the luminous flux in milliamperes per lumen (mA/lm)
>
> S is the area of the light-sensitive surface in square centimeters (cm$^2$ )

### Formula 213.4
### Photoelectric Current

$$If = exE$$

*or*

$$If = sx\Phi$$

Where:

> If is the photoelectric current in milliamperes (mA)
>
> e is the ilumination sensitivity in milliamperes per lux (mA/lx)
>
> E is the luminous intensity in lux (lx)
>
> s is the sensitivity to the luminous flux in milliamperes per lumen (mA/lm)
>
> $\Phi$ is the luminous flux in lux (lx)

# 214. LDR OR PHOTORESISTOR

LDRs (Light-Dependent Resistors), CdS cells (Cadmium-sulfide cells), and photoresistors are devices whose resistance depends on light striking a sensitive surface. The next formulas describe the performance of those devices. *Figure 191* shows a circuit with the parameters used in the formulas and calculations.

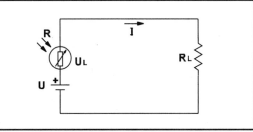

*Figure 191*

# Formula 214.1
# Resistance vs Illumination

$$R = AxL^{-\alpha}$$

Where:

R is the resistance in ohms ($\Omega$)

L is the illumination in lux (lx)

$\alpha$ is a constant given by the manufacturer

A is a constant given by the manufacturer

NOTE: 1.) The value of $\alpha$ depends on the cadmium-sulfide used and on the manufacturing process. Typical values are in the range between 0.6 and 0.9.

2.) The peak of sensitivity at about 680 nm.

3.) LDRs are slow devices with a recovery rate not greater than 200 kohm/s.

The recovery rate is greater in the reverse direction, or when going from darkness to an illumination level.

# Formula 214.2
# Photoelectric Current

$$If = \frac{nxex\tau}{T}$$

Where:

If is the photoelectric current as function of the amount of light, in amperes (A)

n is the number of charge carriers per second

e is the elementary charge in coulombs (C) or amperes x seconds (As)

T is transit time of the charge carriers between electrodes in seconds (s)

e = 1.6 x 10$^{-19}$ C (Coulombs) or As (amperes x seconds)

## Formula 214.3
## Dark Current

$$Id = \frac{U}{Rd}$$

Where:

Id is the dark current in ampares (A)

U is the voltage applied to the device in volts (V)

Rd is the dark resistance in ohms ($\Omega$)

## Formula 214.4
## Dissipation Power

$$P = \frac{U^2}{R}$$

Where:

P dissipated power in watts (W)

U is the voltage applied to the device in volts (V)

R is the resistance in ohms ($\Omega$)

## Formula 214.5
## Maximum Current

$$Imax = \sqrt{\frac{Pmax}{R}}$$

Where:

Imax is the maximum current through the device in amperes (A)

Pmax is the maximum dissipation power in watts (W)

R is the resistance with a given amount of light, in ohms ($\Omega$)

**Application Example**

Calculate the maximum current flowing to an LDR when the resistance falls to 1000 ohms. The LDR maximum dissipation power is 500 mW.

Data:

R = 1000 ohms

Pmax = 500 mW = 0,5 W

Applying Formula 214.5:

$$\mathrm{Im}ax = \sqrt{\frac{0.5}{1000}}$$
$$\mathrm{Im}ax = \sqrt{0.0005}$$
$$\mathrm{Im}ax = 0.022\,A$$

The maximum current is 0.022A or 22 mA.

# 215. PHOTODIODE

A reverse-biased semiconductor diode has a resistance depending on the amount of charge carrier liberated by thermal effect. But charge carriers can also be liberated by an external radiation, such as light. This effect is used in photodiodes, devices with a reversed resistance depending on the amount of light falling onto their junctions. *Figure 192* shows a typical photodiode with its characteristic curve. The next formulas describe the electric performance of these devices.

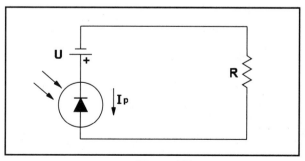

*Figure 192*

## Formula 215.1
### Photocurrent as a Function of Radiant Power

$$Ip = sx\Phi$$

Where:

Ip is the photocurrent in milliamperes or microamperes (mA or $\mu$ A)

s is the spectral photosensitivity in milliamperes per lumen (mA/lm)

$\Phi$ is the radiant power of luminous flux in lumens (lm)

# 216. PHOTOTRANSISTOR

The phototransistor operates as a photodiode. The only difference is that the photocurrent is amplified by its own component. The next formulas are used to describe the performance of this component.

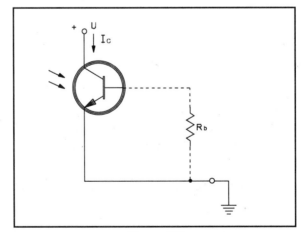

*Figure 193*

## Formula 216.1
### Collector Current

$$Ic = (1 + \beta)xIcbo$$

Where:

Ic is the collector current in amperes (A)

$\beta$ is the transistor gain in the common-emitter configuration

Icbo is the current between base and collector with open emitter, in amperes (A)

## Formula 216.2
## Sensivity to Luminous Flux

$$s = \frac{e}{S}$$

Where:

s is the sensitivity to the luminous flux in milliamperes per lumen (mA/lm)

e is the sensitivity to illumination in milliamperes per lux (mA/lx)

A is the illuminated surface in square centimeters (cm$^2$ )

## Formula 216.3
## Sensitivity to Illumination

$$e = \frac{Ic}{E}$$

Where:

Ic is the collector current in milliamperes (mA)

e is the sensitivity to illumination in milliamperes per lux (mA/lx)

E is the illumination intensity in lux (lx)

## Formula 216.4
## Sensitivity to Luminous Flux

$$s = \frac{Ic}{\Phi}$$

Where:

s is the sensitivity to the luminous flux in milliamperes per lumen (mA/lm)

Ic is the collector current in milliamperes (mA)

$\Phi$ is the luminous flux in lumens (lm)

# COLORIMETRY

## TABLE 72
## Color Primary Valences

| Primary Valences | Wavelength ($\lambda$) in nm | Radiation Relative Density |
|---|---|---|
| Red ($\overline{R}$) | 700.0 | 73.0420 |
| Green ($\overline{G}$) | 546.1 | 1.3971 |
| Blue ($\overline{B}$) | 435.8 | 1.0000 |

## TABLE 73
## Standard Color Valences

$$\overline{X} = 2.36460.\overline{R} - 0.51515.\overline{G}$$
$$\overline{Y} = -89653.\overline{R} + 1.42640.\overline{G} -$$
$$\overline{Z} = -0.46807.\overline{R} + 0.08875.\overline{C}$$

# 217. STEFAN-BOLTZMANN'S LAW

The rate of radiation of energy of all wavelengths from a black body is proportional to the fourth power of the absolute temperature. *Figure 194* shows the spectral energy distribution of an incandescent carbon as a function of the filament temperatures.

# Formula 217.1
# Stefan-Boltzmann's Law

$$\varepsilon = \sigma x T^4$$

Where:

$\varepsilon$ is the rate of radiation of energy

$\sigma$ is the coefficient of proportionality = 5.67 x 10$^{-12}$ w/cm$^2$ $^\circ$K$^4$

T is the temperature in degrees Kelvin ($^\circ$K)

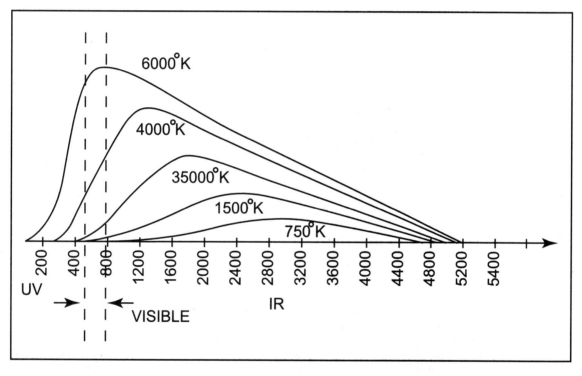

*Figure 194*

# MATHEMATICS

# 218. DIFFERENTIATION FORMULAS

The next are basic formulas used in differentiation.

## Formulas 218.1
## Basic Differentiation

$$\frac{d}{dx}(c) = 0$$

$$\frac{d}{dx(x)} = 1$$

$$\frac{d}{dx}(x^n) = nx^{n-1}$$

$$\frac{d}{dx}(cu) = c\frac{du}{dx}$$

$$\frac{d}{dx}(u+v) = \frac{du}{dx} + \frac{dv}{dx}$$

$$\frac{d}{dx}(u-v) = \frac{du}{dx} - \frac{dv}{dx}$$

$$\frac{d}{dx}(uv) = u\frac{dv}{dx} + v\frac{du}{dx}$$

$$\frac{d}{dx}\left(\frac{u}{v}\right) = \frac{v\frac{du}{dx} + u\frac{dv}{dx}}{v^2}, (v \neq 0)$$

$$\frac{d}{dx}(u^n) = nu^{m-1}\frac{du}{dx}, (u \neq 0)$$

# Formulas 218.2
# Trigonometric Functions

$$\frac{d}{dx}(sinu) = cosu\frac{du}{dx}$$

$$\frac{d}{dx}(cosu) = -sinu\frac{du}{dx}$$

$$\frac{d}{dx}(tgu) = sec^2u\frac{du}{dx}$$

$$\frac{d}{dx}(cotgu) = -cosec^2u\frac{du}{dx}$$

$$\frac{d}{dx}(secu) = secu.tgu\frac{du}{dx}$$

$$\frac{d}{dx}(cosecu) = -coseu.cotgu\frac{du}{dx}$$

## Application Example:

The instantaneous current in a circuit is given by $i = \frac{dq}{dt}$, where q is the charge in Coulombs (C) and t it is the time in seconds (s).

Find i (in amperes) when the current is given by:

$$q = 100t^3 - 20t \qquad \text{for t = 0.5 s}$$

Derivating q = f(t)

$$q = 300t^2 - 20$$

Making t = 0.5:

$$\frac{dq}{dt} = 300x(0.5)^2 - 20$$

$$\frac{dq}{dt} = 300x0.25 - 20$$

$$\frac{dq}{dt} = 75 - 20$$

$$\frac{dq}{dt} = 55$$

$$i = 55 \text{ A}$$

# 219. INTEGRATION FORMULAS

The next are basic integration formulas used in many electronic calculations:

## Formulas 219.1
## Common Integrals

$$\int du = u + C$$

$$\int u^n du = \frac{u^{n+1}}{n+1} + C, (n \neq -1)$$

$$\int k du = k \int du, (k = cons \tan i$$

$$\int (du + dv) = \int du + \int dv$$

$$\int e^u du = e^u + C$$

$$\int \ln u.du = u.\ln|u| - u + C$$

$$\int \frac{du}{u} = \ln|u| + C$$

## Formula 219.2
## Trigonometric

$$\int sinu.du = -\cos u + C$$

$$\int \cos u.du = sinu + C$$

$$\int tgu.du = -\ln(\cos u) + C =$$

$$\int \cot gu.du = \ln(sinu) + C$$

$$\int \sec u.du = \ln(\sec u + tgu) +$$

$$\int \cos ecu.du = \ln(\cos ecu - c$$

## Application Example:

The amount of power produced by a resistor is given by $P = 2t^3$. Determine the medium power dissipated by the resistor in the interval between t = 1 and t = 3 seconds.

$$Pm = \frac{1}{2}\int_{1}^{3} 2t^3 dt$$

$$Pm = \frac{1}{2}\left[\frac{2t^4}{4}\right]_{1}^{3}$$

$$Pm = \frac{1}{2}\left[\frac{81}{2} - \frac{1}{2}\right]$$

$$Pm = \frac{1}{2}x40$$

$$Pm = 20watts$$

## TABLE 74
## Square and Cubic Roots of Some Numbers

Square and cubic roots of numbers up to 100 are common in calculations involving electronics. If the reader doesn't have a calculator in hand the next table will be very useful.

| Number | Square Root | Cubic Root |
|:---:|:---:|:---:|
| 1 | 1.000000 | 1.000000 |
| 2 | 1.414214 | 1.259921 |
| 3 | 1.732051 | 1.442250 |
| 4 | 2.000000 | 1.587401 |
| 5 | 2.236068 | 1.710976 |
| 6 | 2.449490 | 1.817121 |
| 7 | 2.645751 | 1.912931 |
| 8 | 2.828427 | 2.000000 |
| 9 | 3.000000 | 2.080084 |
| 10 | 3.162227 | 2.154435 |
| 11 | 3.316625 | 2.224980 |
| 12 | 3.464102 | 2.289428 |
| 13 | 3.605551 | 2.352335 |
| 14 | 3.741657 | 2.410142 |
| 15 | 3.873983 | 2.466212 |
| 16 | 4.000000 | 2.519842 |
| 17 | 4.123106 | 2.571282 |
| 18 | 4.242640 | 2.620741 |
| 19 | 4.358899 | 2.668402 |
| 20 | 4.472136 | 2.714418 |
| 21 | 4.582576 | 2.758924 |
| 22 | 4.690416 | 2.802039 |
| 23 | 4.795832 | 2.843867 |
| 24 | 4.898980 | 2.884499 |
| (continued next page) | | |

| Number | Square Root | Cubic Root |
|--------|-------------|------------|
| 25 | 5.000000 | 2.944018 |
| 26 | 5.099020 | 2.962496 |
| 27 | 5.196152 | 3.000000 |
| 28 | 5.291502 | 3.036589 |
| 29 | 5.385165 | 3.072317 |
| 30 | 5.477226 | 3.107233 |
| 31 | 5.567764 | 3.141381 |
| 32 | 5.656854 | 3.174802 |
| 33 | 5.744563 | 3.207534 |
| 34 | 5.831952 | 3.239612 |
| 35 | 5.916080 | 3.271066 |
| 36 | 6.000000 | 3.301927 |
| 37 | 6.082763 | 3.332222 |
| 38 | 6.164414 | 3.362975 |
| 39 | 6.245998 | 3.391212 |
| 40 | 6.324555 | 3.420952 |
| 41 | 6.403124 | 3.448217 |
| 42 | 6.480741 | 3.476027 |
| 43 | 6.557438 | 3.503398 |
| 44 | 6.633250 | 3.530348 |
| 45 | 6.708204 | 3.556893 |
| 46 | 6.782330 | 3.583048 |
| 47 | 6.855655 | 3.608826 |
| 48 | 6.928203 | 3.634241 |
| 49 | 7.000000 | 3.659306 |
| 50 | 7.071068 | 3.684031 |
| 51 | 7.141428 | 3.708430 |
| 52 | 7.211102 | 3.732511 |

(continued next page)

| Number | Square Root | Cubic Root |
|:------:|:-----------:|:----------:|
| 53 | 7.280110 | 3.756286 |
| 54 | 7.348469 | 3.779763 |
| 55 | 7.416198 | 3.802953 |
| 56 | 7.483315 | 3.825962 |
| 57 | 7.549834 | 3.848501 |
| 58 | 7.615773 | 3.870877 |
| 59 | 7.681146 | 3.893996 |
| 60 | 7.746967 | 3.914868 |
| 61 | 7.810250 | 3.936497 |
| 62 | 7.874008 | 3.977892 |
| 63 | 7.937254 | 3.979057 |
| 64 | 8.000000 | 4.000000 |
| 65 | 8.062258 | 4.020726 |
| 66 | 8.124039 | 4.041240 |
| 67 | 8.185352 | 4.061548 |
| 68 | 8.246211 | 4.981655 |
| 69 | 8.306623 | 4.101566 |
| 70 | 8.366600 | 4.121285 |
| 71 | 8.426149 | 4.140818 |
| 72 | 8.485281 | 4.160168 |
| 73 | 8.544003 | 4.179339 |
| 74 | 8.602325 | 4.198337 |
| 75 | 8.660254 | 4.217164 |
| 76 | 8.717798 | 4.235824 |
| 77 | 8.775964 | 4.254321 |
| 78 | 8.831760 | 4.272659 |
| 79 | 8.888194 | 4.290841 |
| 80 | 8.944272 | 4.308869 |

(continued next page)

| Number | Square Root | Cubic Root |
|--------|-------------|------------|
| 81 | 9.000000 | 4.326749 |
| 82 | 9.055386 | 4.344481 |
| 83 | 9.110434 | 4.362071 |
| 84 | 9.165152 | 4.379519 |
| 85 | 9.219544 | 4.396830 |
| 86 | 9.273619 | 4.414005 |
| 87 | 9.327379 | 4.431047 |
| 88 | 9.380832 | 4.448960 |
| 89 | 9.434981 | 4.464745 |
| 90 | 9.486833 | 4.481405 |
| 91 | 9.539392 | 4.497941 |
| 92 | 9.591663 | 4.514358 |
| 93 | 9.643651 | 4.530655 |
| 94 | 9.695360 | 4.546836 |
| 95 | 9.746795 | 4.562902 |
| 96 | 9.798959 | 4.578857 |
| 97 | 9.848858 | 4.594701 |
| 98 | 9.890495 | 4.610436 |
| 99 | 9.949874 | 4.626065 |
| 100 | 10.000000 | 4.641589 |

# 220. TOLERANCES

The difference between the real value and the specified value of a device is called tolerance and can be expressed in terms of percentage. The next formula is used to calculate tolerance.

**Formula 220.1**
**Tolerance**

$$T = \frac{a}{N} x100$$

Where:

> T is the tolerance in percent (%)
>
> a is the real value (any unit)
>
> N is the specified value (any unit)

## Derivated Formulas:

## Formula 220.2
## Range of Values

$$a = \frac{TxN}{100}$$

Where:

> a is the variation of value of the the device
>
> T is the tolerance in percent (%)
>
> N is the specified value

### Application Example:

A 200-ohm x 10% resistor is used in a project. What is the variation of value that can be expected of this component?

Data:

> N = 200 ohms
>
> T = 10%
>
> a = ?

Using Formula:

$$a = \frac{200x10}{100}$$
$$a = 20 ohms$$

## TABLE 75
## Laplace Transforms

Anyone familiar with the necessary calculus can evaluate the Laplace transforms of any function by performing the necessary operations. However, the process for common functions is tedious, and is unnecessary when precalculated tables are available. Next is a table of common transforms involving electronics.

|   | Time Function | Laplace Transform |
|---|---------------|-------------------|
| 1 | 1 (unit step) | $\dfrac{1}{s}$ |
| 2 | t (ramp) | $\dfrac{1}{s^2}$ |
| 3 | $e^{at}$ | $\dfrac{1}{s-a}$ |
| 4 | $e^{-at}$ | $\dfrac{1}{s+a}$ |
| 5 | $te^{at}$ | $\dfrac{1}{(s-a)^2}$ |
| 6 | $te^{-at}$ | $\dfrac{1}{(s+a)^2}$ |
| 7 | $\dfrac{t^{n-1}}{(n-1)!}$ | $\dfrac{1}{s^n}, n = 1,2,3...$ |
| 8 | $sinkt$ | $\dfrac{k}{s^2+k^2}$ |
| 9 | $e^{at}sinkt$ | $\dfrac{k}{(s-a)^2+k^2}$ |
| 10 | $coskt$ | $\dfrac{s}{s^2+k^2}$ |
| | (continued next page) | |

| | Time Function | Laplace Transform |
|---|---|---|
| 11 | $e^{at}\cos kt$ | $\dfrac{s-a}{(s-a)^2+k^2}$ |
| 12 | $e^{at}-e^{bt}$ | $\dfrac{a-b}{(s-a)(s-b)}$ |
| 13 | $1-e^{at}$ | $\dfrac{-a}{s(s-a)}$ |
| 14 | $ae^{at}-be^{bt}$ | $\dfrac{s(a-b)}{(s-a)(s-b)}$ |
| 15 | $\dfrac{t^{n-1}e^{at}}{(n-1)!}$ | $\dfrac{1}{(s-a)^n}, n=1,2,3....$ |
| 16 | $e^{at}(1+at)$ | $\dfrac{s}{(s-a)^2}$ |
| 17 | $t.\sin kt$ | $\dfrac{2ks}{(s^2+k^2)^2}$ |
| 18 | $t.\cos kt$ | $\dfrac{s^2-k^2}{(s^2+k^2)^2}$ |
| 19 | $\sin kt - kt\cos t$ | $\dfrac{2k^2}{(s^2+k^2)^2}$ |
| 20 | $\sin kt + \cos kt$ | $\dfrac{2ks^2}{(s^2+k^2)^2}$ |
| 21 | $1-\cos kt$ | $\dfrac{k^2 s}{s(s^2+k^2)}$ |
| 22 | $kt - \sin kt$ | $\dfrac{k^2}{s^2(s^2+k^2)}$ |

# 221. FOURIER TRANSFORMS

The Fourier's theorem states that it is always possible to select numbers from **a1** to **an** in the following series and $\varphi_1$ to $\varphi_n$ in such a manner that any periodic vibration having the frequency f ($\omega t$) may be represented in the form of a sum of harmonic vibrations.

## Formula 221.1
## Fourier's Theorem

$$x = a_1 \cos(\omega t + \varphi_1) + a_2 \cos(2\omega t + \varphi_2) + a_3 \cos(3\omega t + \varphi_3 + \ldots\ldots + a_n \cos(n\omega t + \varphi_n)$$

Where:

x is the instantenous amplitude of the wave

a1 to an are the amplitudes of the harmonics

$2\omega, 3\omega, \ldots n\omega$ are the overtones or harmonics

$\varphi_1, \varphi_2, \ldots \varphi_n$ are phase angles

In the next tables we give the harmonic composition of comon waveforms determined by the Fourier Transform. The corresponding percentage is also given to each component (harmonic) making the calculations easier.

# TABLE 76
## Square Wave
## Harmonic Composition

| Harmonic | Relative Magnitude | Percentual Magnitude (%) |
|---|---|---|
| Fundamental | $\dfrac{4}{\pi}U$ | 127 |
| 2 nd | 0 | 0 |
| 3 rd | $\dfrac{4}{3\pi}U$ | 42.5 |
| 4 th | 0 | 0 |
| 5 th | $\dfrac{4}{5\pi}U$ | 25.5 |
| 6 th | 0 | 0 |
| 7 th | $\dfrac{4}{7\pi}U$ | 18.2 |

# TABLE 77
## Triangular Wave
## Harmonic Composition

| Harmonic | Relative Magnitude | Percentual Magnitude (%) |
|---|---|---|
| Fundamental | $\dfrac{8}{\pi^2}U$ | 81 |
| 2 nd | 0 | 0 |
| 3 rd | $\dfrac{8}{9\pi^2}U$ | 9 |
| 4 th | 0 | 0 |
| 5 th | $\dfrac{8}{25\pi^2}U$ | 3.2 |
| 6 th | 0 | 0 |
| 7 th | $\dfrac{8}{49\pi^2}U$ | 1.6 |

# TABLE 78
## Sawtooth Wave
## Harmonic Composition

| Harmonic | Relative Magnitude | Percentual Magnitude (%) |
|---|---|---|
| Fundamental | $\dfrac{2}{\pi}U$ | 63.6 |
| 2 nd | $\dfrac{1}{\pi}U$ | 31.8 |
| 3 rd | $\dfrac{2}{3\pi}U$ | 21.2 |
| 4 th | $\dfrac{1}{2\pi}U$ | 15.9 |
| 5 th | $\dfrac{2}{5\pi}U$ | 12.7 |
| 6 th | $\dfrac{1}{3\pi}U$ | 10.6 |
| 7 th | $\dfrac{2}{7\pi}U$ | 9.1 |

**TABLE 79**

**Half-Wave Rectifier Waveform**

**Harmonic Composition**

| Harmonic | Relative Magnitude | Percentual Magnitude (%) |
|----------|--------------------|--------------------------|
| Fundamental | $\dfrac{1}{\pi}U$ | 31.8 |
| 2 nd | $\dfrac{2}{3\pi}U$ | 21.2 |
| 3 rd | 0 | 0 |
| 4 th | $\dfrac{2}{15\pi}U$ | 4.2 |
| 5 th | 0 | 0 |
| 6 th | $\dfrac{2}{35\pi}U$ | 1.8 |
| 7 th | 0 | 0 |

# TABLE 80

## Full-Wave Rectifier Waveform

## Harmonic Composition

| Harmonic | Relative Magnitude | Percentual Magnitude (%) |
|----------|--------------------|--------------------------|
| Fudamental | $\dfrac{2}{\pi}U$ | 63.6 |
| 2 nd | $\dfrac{4}{3\pi}U$ | 42.3 |
| 3 rd | 0 | 0 |
| 4 th | $\dfrac{4}{15\pi}U$ | 8.5 |
| 5 th | 0 | 0 |
| 6 th | $\dfrac{4}{35\pi}U$ | 3.6 |
| 7 th | 0 | 0 |

## TABLE 81
## Physical Constants

| Constant | Symbol | Value | Unit |
|---|---|---|---|
| Speed of light in vacuum | c | $2.997\ 925\ 0 \times 10^8$ | m/s |
| Elementary charge | e | $1.602\ 191\ 7 \times 10^{-19}$ | C |
| Avogadro constant | N | $6.022\ 169 \times 10^{23}$ | $mol^{-1}$ |
| Atomic mass unit | u | $1.660\ 531 \times 10^{-27}$ | kg |
| Electron rest mass | $m_e$ | $9.109\ 558 \times 10^{-31}$ | kg |
| Proton rest mass | $m_p$ | $1.672\ 614 \times 10^{-27}$ | kg |
| Faraday constant | F | $9.648\ 670 \times 10^4$ | C/mol |
| Planck constant | h | $6.626\ 196 \times 10^{-34}$ | J.s |
| Fine structure constant | $\alpha$ | $7.297\ 759$ | - |
| Rydeberg constant | $R\infty$ | $1.097\ 373\ 12 \times 10^7$ | $m^{-1}$ |
| Bohr Magnetron | $\mu B$ | $9.274\ 096 \times 10^{-24}$ | J/T |
| Boltzmann constant | k | $1.380\ 622 \times 10^{-23}$ | J/K |
| Gravitational constant | G | $6.673\ 2 \times 10^{-11}$ | $N.m^2/kg^2$ |

# REFERENCES

1 - *2000 Transistores FET*, Fernando Estrada, Saber, Brazil, 1989

2 - *Altavoces y Cajas de Resonancia Para Hifi* - H. H. Klinger, Marcombo, Barcelona, 1976

3 - *Building Speaker Systems*, M. D. Hull, Philips, 1977

4 - *Circuitos e Informações*, Newton C. Braga, Editora Saber,Brazil, 1979

5 - *CMOS Cookbook*, Don Lancaster, H. W. Sams, 1980

6 - *CMOS Logic Databook*, National Semniconductor, 1988

7 - *Curso Básico de Eletrônica*, Newton C. Braga, Editora Saber, Brazil, 1996/1999

8 - *Dix Enceintes Acoustiques HI-FI*, Pierre Chauvigni, Editions Radio, Paris, 1976

9 - *Electronic Databook*, Rudolf F. Graf -Van Nostrand Reihold Co., 1974

10 - *Electrónica Digital Moderna*, J. M. Angulo, Paraninfo, Spain, 1986

11 - *Electronica e Musica Pop*, Hans Goddin, paraninfo, Spain, 1982

12 - *Electronica, Formulas, Problemas, Tablas, Circuitos Integrados*, Alfredo Borque, Paraninfo, Spain, 1983

13 - *Enciclopédia Moderna de Electronica, Tablas, Graficos, Códigos, Datos Utiles*, Radiorama, Argentina, 1972

14 - *Formulário de Electronica*, Francisco Ruiz Vassallo, CEAC, Spain. 1988

15 - *Formulas y Calculos Para Electronica y Radio*, W.E. Pannett, Paraninfo, Madrid, 1963

16 - *Fourier Analysis*, Hwei P. Hsu, Simon & Shuster Inc. 1972

17 - *Handbook of Electronic Tables*, Martin Clifford, Tab Books, 1972

18 - *Handbook of Elementary Physiscs*, N. Koshkin and M. Shirkevich, MIR, Moscow, 1968

19 - *Hearing Aid Handbook*, Wayne J. Staab, Tab Books, 1978

20 - *How to Design and Build Audio Amplifiers*, Mannie Horowitz, Tab Books 1980

21 - *Impedance*, Rufus P. Turner, Tab Books 1976

22 - *Introdução À Eletrônica*, Wilson José Tucci, Nobel, Brazil, 1979

23 - *Linear Applications Handbook*, National Semiconductor, 1991

24 - *Linear Circuits Data Book*, Texas Instruments, 1984

25 - *Manual de Instruments de Medida Eléctronicos*, Francisco Luiz Vassalo, CEAC, Barcelona, Spain, 1981

26 - *Op Amp Circuit Design and Applications*, Joseph Carr, Tab Books, 1976

27 - *Optics*, Francis Weston Sears, Addison Wesley Publishing Co. 1958

28 - *Optotoelectronics Theory and Practice*, Alan Chappell, Texas Instruments, 1976

29 - *Practical Circuit Design For The Experimenter*, Don Tuite, TAB Books, 1974

30 - *Practical Electronics Math*, Tab Books, 1982

31 - *Practical Triac/SCR Projects For The Experimenter*, R.W. Fox, Tab Books 1974

32 - *Professional Electrical/Electronic Engineer's Licence Study Guide*, Edward J. Ross , TAB Books, 1977

33 - *Radio Praktiker Bücherol Kleine Elektronik-Formelsammlung*, Georg Rose, Franzis-Verlog München, 1976

34 - *Radiotron Designer's Handbook*, RCA - Langford-Smith-1953

35 - RCA Databook, *CMOS Integrated Circuits*, 1983

36 - *Sourcebook of Electronic Organ Circuits*, Alan Douglas and S. Astley, Tab Books, 1975

37 - *Technical Calculus*, Da le Ewen and Michael A. Topper, Prentice Hall Inc., 1977

38 - *Technische Formelsammlung*, Kurt Gieck, Gieck Verlag, West Germany, 1979

39 - *Teoria e Desenvolvimento de Projetos de Circuitos Eletrônicos*, Antonio Marco V. Cipelli and Waldir João Sandrini, Erica, Brazil

40 - *The Power Semiconductor Data Book*, Texas Instruments, 1972

41 - *Transistors in Radio, Television and Electronics*, Milton S. Kiver, McGraw-Hill Book Company, 1959

42 - *TTL Cookbook*, Don Lancaster, H. W. Sams, 1977

43 - *Varistors, Thermistors and Sensors*, Philips Components, 1989

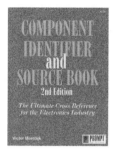

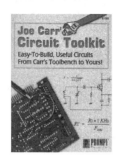

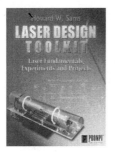